T0269466

SpringerBriefs in Physics

More information about this series at http://www.springer.com/series/8902

SpringerBriefs in Physics

More information about this series at http://www.springer.com/series/8902

Arne Grenzebach

The Shadow of Black Holes

An Analytic Description

Arne Grenzebach
ZARM—Zentrum für angewandte
Raumfahrttechnologie
und Mikrogravitation
Universität Bremen
Bremen
Germany

ISSN 2191-5423 ISSN 2191-5431 (electronic)
SpringerBriefs in Physics
ISBN 978-3-319-30065-8 ISBN 978-3-319-30066-5 (eBook)
DOI 10.1007/978-3-319-30066-5

Library of Congress Control Number: 2016936995

Printed on acid-free paper

This Springer imprint is published by Springer Nature
The registered company is Springer International Publishing AG Switzerland

Phantasie ist wichtiger als Wissen,
denn Wissen ist begrenzt.

Albert Einstein

Preface

All the work presented here covers an analytic geometrical way to construct the shadow of black holes. The shape of the shadow varies in different space-times, i.e., it depends on specific properties of the black hole as, for example, the spin. My aim is to provide calculations as general as possible.

This short book summarizes the scientific results of my doctoral project where I generalized the existing calculations for the shadow of a Kerr black hole. I found analytical formulas for the boundary of the shadow for the general Plebański–Demiański class of stationary, axially symmetric type D solutions of the Einstein–Maxwell equations. As far as I know, such formulas did not exist before not even in the Kerr space-time. With my formulas, it is possible to calculate the shadow for observers at arbitrary positions. In addition, the shadow-plots can be compared with those of a moving observer. If the motion of the observer is in purely radial direction, then the aberration formula of Penrose is recovered from my formulas.

As pointed out in Chap. 1, the existence of the photon region is crucial for determining the shadow of a black hole. This results in the following natural structure of this thesis. In Chap. 2, I discuss in some detail the Plebański–Demiański class of space-times and review relevant properties of its metric. The geometrically important photon region and other interesting regions in the environment of a black hole are considered in Chap. 3. The last chapter, Chap. 4, is dedicated to deduce the formulas that describe the boundary curve of the black hole's shadow.

Large parts of the scientific results are already published in three papers. The corresponding paragraphs are marked in the following references which refer to my papers Grenzebach et al. (2014, 2015), Grenzebach (2015), respectively. Sentences marked with [*i*] can be found in total or only slightly modified in the *i*th paper.

Bremen, Germany
January 2016

Arne Grenzebach

References

Grenzebach A, Perlick V, Lämmerzahl C (2014) Photon regions and shadows of Kerr–Newman NUT Black Holes with a cosmological constant. Phys Rev D 89:124,004(12). doi:10.1103/PhysRevD.89.124004, arXiv:1403.5234

Grenzebach A (2015) Aberrational effects for shadows of black holes. In: Puetzfeld et al. pp 823–832. doi:10.1007/978-3-319-18335-0_25, proceedings of the 524th WE-Heraeus-Seminar "Equations of Motion in Relativistic Gravity", Bad Honnef, Germany, 17–23 February 2013, arXiv:1502.02861

Grenzebach A, Perlick V, Lämmerzahl C (2015) Photon regions and shadows of accelerated black holes. Int J Modern Phys D 24(9):1542,024(22). doi:10.1142/S0218271815420249, "Special Issue Papers" of the "7th Black Holes Workshop", Aveiro, Portugal. arXiv:1503.03036

Acknowledgements

During the whole doctoral project, I got a lot of support by several people whom I would like to thank all.

First of all, I want to thank my supervisor Claus Lämmerzahl for providing me with this wonderful and interesting topic, his advice and for introducing me to several scientists. I am indebted to Volker Perlick for many interesting and inspiriting discussions. Without his supervision, my thesis would not exist. Furthermore, I would like to thank Jutta Kunz-Drolshagen for acting as the second referee. I owe thank to Olaf Lechtenfeld for being my second supervisor within the research training group and for his kind financial support while writing the thesis.

I am grateful for many fruitful discussions with Nico Giulini, Saskia Grunau, Norman Gürlebeck, Eva Hackmann, David Kofron, Jutta Kunz-Drolshagen, Luciano Rezzolla and Ziri Younsi. Special thanks go to Norman Gürlebeck, Eva Hackmann, Sven Herrman, Sascha Kulas, Christian Pfeifer, Dennis Philipp, Christian Vogt and Ziri Younsi for reading my manuscript and pointing out several weak points. In particular, the greatest thanks go to my brothers Claas and Gerrit for their wonderful support and help with the last things.

Likewise, I would like to thank all my colleagues of the fundamental physics group at ZARM for their support and the good time. Thanks go to my colleagues of "Models of Gravity". Furthermore, I thank Luciano Rezzolla and his group in Frankfurt for the warm hospitality during my research stay.

Additionally, I want to express my gratitude to my office colleagues, Andreas Resch and Christian Vogt for all these years of kind support.

The WE-Heraeus foundation deserves my gratitudes for offering poster awards. I gratefully acknowledge support from the DFG within the Research Training Group 1620 "Models of Gravity" and from the "Centre for Quantum Engineering and Space-Time Research (QUEST)".

I would like to thank Sabine Lehr and Abirami Purushothaman (Springer Science+Business Media) for their support during the publishing process.

For some time, the instruments of Max and Heinrich Thein give me a lot of pleasure. I would like to thank both, their craftsman as well as Uli Beckerhoff, Hans Kämper, Matthias Wulff, Evgeny Yatsuk and all other musicians.

Finally, I am deeply grateful to my parents and my brothers Gerrit and Claas with Claudia for their constant support, encouragement and believing in me over all these years. I thank Christina and Philipp Niemann for their friendship.

Contents

List of Figures

List of Tables

Abstract

With the upcoming high-resolution observations of the Galactic center, it will be revealed whether our Milky Way hosts a black hole in its center. Due to the strong gravity and the resulting deflection of light, the black hole will cast a shadow and an observed image of the shadow would be a strong evidence for the existence of black holes. It is expected that the *Event Horizon Telescope* or the *BlackHoleCam* project will produce a radio image of the shadow of the central black hole in a few years' time. Therefore, it is about time to advance the theoretical investigations of the shadows of black holes as far as possible, as a basis for evaluating the observational results.

This short book is about an analytic way to describe the shadow of black holes. As an introduction, I present a survey of the attempts to observe the shadow of the black holes in our Galaxy near Sgr A* and in the neighbouring galaxy M87. Black holes are described by metrics of the general Plebański–Demiański class of space-times. All these metrics are axially symmetric and stationary type D solutions to the Einstein–Maxwell equations with a cosmological constant. The space-times are characterized by seven parameters: mass, spin, electric and magnetic charge, gravitomagnetic NUT charge, a so-called acceleration parameter and the cosmological constant.

Based on a detailed discussion of the metrics, I derive analytical formulas for the photon regions (regions that contain spherical lightlike geodesics) and for the boundary curve of the shadow as it is seen by an observer at the given Boyer–Lindquist coordinates in the domain of outer communication. They enable me to analyze the dependency of the shadow of a Kerr black hole on the motion of the observer. For all cases, the photon regions and shadows are visualized for various values of the parameters. The analytical formulas are used to find explicit expressions for the horizontal and vertical angular diameters of the shadow. Finally, these values are estimated for the black holes at the center of our Galaxy and of M87.

Chapter 1
Introduction

Abstract Black holes are intriguing astrophysical objects. But it is still unproven whether black holes exists. Therefore, the observational evidence for black holes is discussed. This is followed by a survey of the attempts to observe the shadow of the black holes in our Galaxy near Sagittarius A* and in the neighbouring galaxy M87 by the European BlackHoleCam project and the US-led Event Horizon Telescope project.

Keywords Observational evidence black holes · Shadow observation · Galactic center · Observing Sgr A* · Observing M87 · Black hole cam · Event horizon telescope · VLBI observation

Black holes are perhaps the most fascinating objects in Astrophysics. It is hard to find any other object or topic that attracts more attention. There are comics that explain black holes (Petit 1995) or movies where a black hole plays a prominent role like the recently released Hollywood movie *Interstellar*. In the movie, it is shown how nearby observers see the shadow of a rotating black hole surrounded by an accretion disk (James et al. 2015).[3]

This optical phenomenon arises because light in the gravitational field of a black hole propagates along curved lines instead of straight lines. Actually, the light deflection is so strong that spherical light paths exist—this region is called *photon region*—and even an event horizon. Whatever passes the horizon is captured for evermore, even light. But already anything coming from outside and crossing the photon region has to pass the event horizon. Consequently, a black hole satisfies its name and an observer really sees just a black spot (as long as there are no light sources in between). This black spot is called the *shadow* of the black hole; it could be considered as an image of the photon region.

In this thesis, we calculate for a general class of space-times what the shadow of black holes looks like. The theoretical basis is of course Albert Einstein's theory of general relativity (Einstein 1915a, b, c) where extensive differential geometric skills are required for the understanding of the theory and their physical consequences. But for our calculations of the shadow we do not need much more than the mathematical description of space-times which are solutions of Einstein's field equations. Consequently, the curvature of the space-time is determined by the matter which in turn defines the motion of particles or light.

A. Grenzebach, *The Shadow of Black Holes*,
SpringerBriefs in Physics, DOI 10.1007/978-3-319-30066-5_1

Now, black holes are special solutions with increasing curvature while approaching the black hole; the curvature diverges in the center of the black hole because of a singularity. The simplest solution that describes a spherically symmetric, static, uncharged black hole is the *Schwarzschild* metric (Schwarzschild 1916); the radius of its event horizon is named the *Schwarzschild radius*. Rotating black holes are described by the *Kerr* metric (Kerr 1963).

The first confirmation of the relativistic light deflection results from observations of positions of stars during the total solar eclipse on May, 29th in the year 1919 (Dyson et al. 1920; Kennefick 2009). In two expeditions, which took the astronomer Arthur Stanley Eddington and Edwin Turner Cottingham to the Island of Príncipe (near western Africa) as well as Charles Davidson and Andrew Crommelin to Sobral (Brazil), the positions of stars near the limb of the covered sun were compared with recorded positions. The resulting deviation matches the relativistic prediction of Einstein (1916) which is twice as big as the classical Newtonian value which was calculated by Henry Cavendish and later by Johann von Soldner (von Soldner 1804; Will 1988).

1.1 Observational Evidence for Black Holes

Since the optical measurements of Cavendish and von Soldner, tremendous progress in astronomical observations has been made. Observations show that galaxies could radiate enormously from a compact central region. A good explanation of these *active galactic nuclei* (AGN) is the accretion of matter onto a supermassive black hole (Lynden-Bell 1969; Lynden-Bell and Rees 1971; Rees 1974, 1984; Müller 2004) where the matter comes from an *accretion disk* accumulated around the galaxies' center. The accretion model is supported by observations of Tanaka et al. (1995). They found that the Fe Kα line in the X-ray emission from ionized iron in the galaxy MCG–6–30–15 is extremely broad and redshifted which indicates that the radiation has to be emitted very close (3 to 10 Schwarzschild radii) to the innermost region. Powered by accretion, matter could be ejected from the galaxies as enormous jets of several thousand light years as it is the case in the elliptical galaxy Messier 87 (M87) shown in Fig. 1.1b. It is ongoing work to examine the exact formation mechanisms of these jets (Doeleman et al. 2012).

Besides the supermassive black holes (10^6–10^{10} solar masses $M_\odot$), which are predicted to be at the centers of most—if not all—galaxies, there are stellar mass black holes. These are remnants of stars after supernovae explosions. They are assumed to be formed when very high mass stars collapse after fusing all nuclei to iron.

Also our Milky Way hosts a powerful radio source in its dynamical center; the source was discovered 1974 and named *Sagittarius A* (Sgr A*)* because it is located in the constellation Sagittarius (Balick and Brown 1974; Goss et al. 2003), see Fig. 1.1a for an infrared image of the central arcseconds of the Galactic center. But not only the radio source Sgr A* was observed. It follows from the rise-and-decay times of flare events that the emission region is small compared to an accretion disk (Gillessen et al.

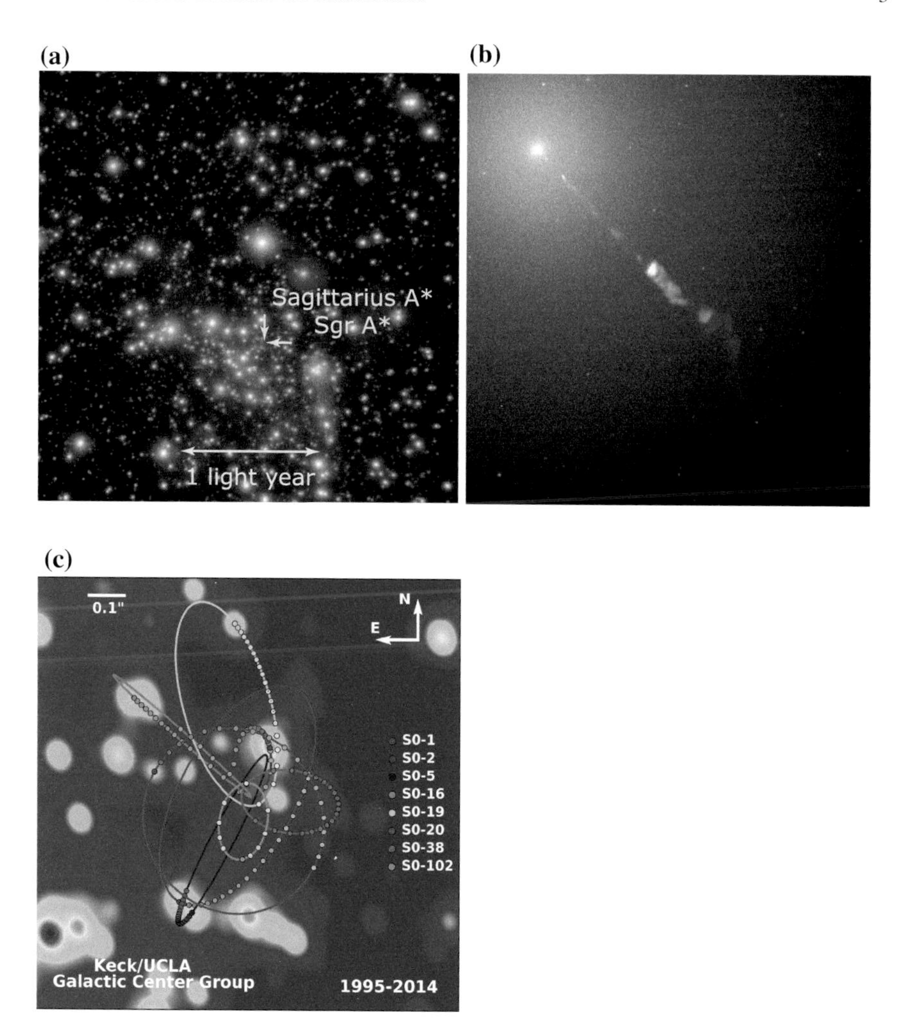

Fig. 1.1 Images of the Galactic center in our Milky Way, of stars orbiting Sagittarius A* in the Galactic center and of the supergiant elliptical galaxy Messier 87 with its jet. **a** Galactic center with Sagittarius A*. **b** Galaxy Messier 87 (M87) with jet. **c** Stellar orbits in the central arcseconds of the Galactic center ("The annual average positions for these stars are plotted as *colored dots*, which have increasing color saturation with time," © Ghez). *Credits*. **a** original image https://www.eso.org/public/images/eso0846a/ © ESO/S. Gillessen et al. **b** original image https://www.spacetelescope.org/images/opo0020a/ © The Hubble Heritage Team (STScI/ AURA) and NASA/ESA, **c** original image http://www.astro.ucla.edu/~ghezgroup/gc/images/research/2014plot_central_image_hires.png © This image was created by Prof. Andrea Ghez and her research team at UCLA and is from data sets obtained with the W.M. Keck Telescopes

2006; Schödel et al. 2005). Additionally, there is strong evidence for the existence of a supermassive black hole associated with the radio source Sgr A*. The evidence results from infrared observations started two decades ago. They show stars orbiting a common central object close to Sgr A* on Keplerian orbits, see Fig. 1.1c, which demonstrates that there has to be a heavy object in a distance of roughly $r_O = 8\,\mathrm{kpc}$[1] with a mass of approximately 4 million solar masses $M_\odot$. Since the mass must be concentrated within a small volume, the most convincing candidate for such an object is a black hole. Latest observations reveal values of $M = 4.31 \times 10^6 M_\odot$ and $r_O = 8.33\,\mathrm{kpc}$ (Gillessen et al. 2009) or $M = 4.1 \times 10^9 M_\odot$ and $r_O = 7.7\,\mathrm{kpc}$ (Meyer et al. 2012a). Older results are given in Eckart and Genzel (1996, 1997), Eisenhauer et al. (2003), Ghez et al. (1998, 2005, 2008). A geometric way how to determine the distance r_O is described by Salim and Gould (1999).

But the observed phenomena are not exclusively explainable by black holes. Alternative models are *gravastars* (Mazur and Mottola 2001) or *holostars* (Petri 2003a, b). Both are static solutions of Einstein's field equations where the outer part is in consequence of Birkhoff's theorem described by the Schwarzschild solution. Their interior solutions differ: gravastars contain a shell of some ultra-relativistic plasma which is stabilized by some type of dark energy while holostars contain a curvature singularity similar to black holes which can be substituted by some sort of quantum object, e.g., a string. But their distinctive feature compared to black holes is the missing event horizon.

By now, observations with higher resolution are possible by combining several radio telescopes via computers to a giant earth-spanning virtual telescope. More details about this observational method called *very-long-baseline interferometry* (VLBI) will be given in Sect. 1.2. Here, we anticipate an exceedingly remarkable result. In a first article, Broderick and Narayan (2006) explain that the observed radiation emitted from Sgr A* and not from the accretion process is much less than one would expect from thermal black body radiation of a surface. Consequently, Sgr A* could not have a surface which is why it has to be a black hole. Broderick's follow-up paper, which takes more recent measurements into account, is more precise: Based on the three assumptions that Sgr A* is gravitationally powered, has reached an approximate steady state and that a presumed surface could be modeled by a thermal spectrum, Broderick et al. (2009) conclude:

> "Recent infrared and millimeter-VLBI observations imply that if the matter accreting onto Sgr A* comes to rest in a region visible to distant observers, the luminosity associated with the surface emission from this region satisfies $L_{\mathrm{surf}}/L_{\mathrm{acc}} \lesssim 0.004$. Equivalently, these observations require that 99.6 % of the gravitational binding energy liberated during infall is radiated in some form prior to finally settling. These numbers are inconsistent by orders of magnitude with our present understanding of the radiative properties of Sgr A*'s accretion

[1] A parsec (abbr. pc, short for *parallax* of one *arcsecond*) is an astronomical unit to measure distances. It is defined as the distance at which the distance Earth–Sun appears at an angle of 1 arcsecond (as). One finds $1\,\mathrm{pc} = 3.09 \times 10^{16}\,\mathrm{m} = 3.26\,\mathrm{ly}$.

flow specifically and relativistic accretion flows generally. Therefore, it is all but certain that no such surface can be present, i.e., *an event horizon must exist*."

An absolutely certain verification of the existence of black holes would be the detection of Hawking radiation, a pure thermal radiation arising from quantum effects (Hawking 1974, 1975, 1976). Unfortunately, for astrophysical objects the radiation is much too weak for being detectable. Nevertheless, the question whether Sgr A* is a black hole or some other object with a surface could be answered at its best with the planned (see Sect. 1.2) high-resolution explorations (in the order of magnitude of the Schwarzschild radius of the central mass) of the center of our galaxy with radio telescopes in submillimeter wavelength range. For black holes, these observations will yield an image of its shadow. But it is still discussed whether imaging a shadow is already sufficient for a proof of the existence of black holes. Cardoso et al. (2014) analyze how background fluctuations influence the stability of ultracompact objects. For rotating stars, already linear fluctuations should cause a fragmentation of the outer layers of the star. Due to the resulting gravitational radiation, the star loses mass and compactness which leads to stable stars without a photon region or even a black hole. Thus, they reasoned that an image of a photon sphere, i.e., a shadow is evidence enough for the existence of black holes!

At this point, it is worth mentioning several informative references. Reviews about the Galactic center research with arguments for the existence of a black hole out there were given by Genzel (2014), Falcke and Markoff (2013), Morris et al. (2012), Genzel et al. (2010), Falcke and Hehl (2003), Melia and Falcke (2001). Observational techniques needed for exploring the Galactic center are described in the book by Eckart et al. (2005) together with historical remarks. Since the whole topic is very popular, there are also several articles in all major science magazines or even in newspapers like the New York Times. As an introduction, we recommend the essays by Melia (2003), Broderick and Loeb (2009), Kruesi[2] (2012) and Britzen (2012).

1.2 Observing the Shadow of Black Holes

Currently, there are two cooperating projects—the US-led *Event Horizon Telescope* (EHT) project[3] and the European *BlackHoleCam* (BHC) project[4]—which are attempting to image the shadow of the black hole at the center of our Galaxy using the very-long-baseline interferometry (VLBI). A detailed explanation how interferometry works and how it is used in radio astronomy is given in the book by Thompson et al. (2004). We just give a brief description.

---

[2]2013 awarded with the David N. Schramm Award for high-energy astrophysics science journalism.

[3]Project website: www.EventHorizonTelescope.org.

[4]Project website: BlackHoleCam.org.

The angular resolution $\delta = \lambda/D$ of a single radio telescope is determined by its aperture D and by the observing wavelength λ. Hence, the resolution can be increased either by observing at shorter wavelengths or by increasing the aperture of the telescope. The latter is limited by material properties which restrict the dish size of fully steerable radio telescopes to a maximum size of about 100 m in diameter. However, it is possible to add up the phase correlated signals of several smaller telescopes from one site; then, the station operates effectively like one single telescope with bigger aperture. Furthermore, the data of several sites of telescopes can be combined. For that purpose, the detected signals are stored and stamped with the exact time of detection determined by ultraprecise atomic clocks. Then, all raw data packets are centrally processed by a supercomputer. Based on the time-stamps, interference patterns between each two telescopes within the network are calculated. With this interferometric method, one can achieve an angular resolution of[5]

$$\delta = \frac{\lambda}{D} = \frac{c}{\nu D} \tag{1.1}$$

where D is now the longest *baseline*, i.e., the longest distance between two telescope stations within the network! Thus, a virtual telescope of the size of the whole array is built and this is why the method is called *very-long-baseline interferometry* (VLBI).

For the EHT or BHC project, the array has Earth-spanning scale with telescopes at several widespread locations around the Earth, see Table 1.1. ALMA[6] and the South Pole Telescope (SPT) will join the network soon which will improve the resolution. The expanded array features baselines of about 10000 km length (cf. Inoue et al. 2014, Table 1) which is 78 % of the Earth's diameter of 12742 km. In Table 1.2, achievable angular resolutions for a baseline of 10000 km for typical millimeter and submillimeter wavelengths are collected.

Based on the interference patterns, the brightness of the source, i.e., an image of the observed sky can be reconstructed, see Thompson et al. (2004), Thiébaut (2009). As a matter of fact, each baseline represents a grid point of the approximation of the (complex) visibility function $V(u, v)$ that is the 2D Fourier transform of the brightness. In order to improve the approximation of V, it is fortunately not mandatory to expand the telescope array (which would not be affordable). The Earth's rotation helps instead. In consequence, all telescopes are rotated relative to the source which yields rotated grid points. Thus, (under the assumption of a sufficiently stable visibility function) just simple multiple repeated measurements enhance the image quality. In the literature, the rotation of the Earth can be identified by bows that are drawn in "u-v coverages". Nevertheless, the reconstructed image is defective, since V is only given at a discrete number of points. But the quality of this "dirty" image can be significantly improved by image processing methods such as deconvolution.

[5]Wavelength λ and frequency ν are linked by the speed c of light: $\lambda\nu = c = 299\,792\,458\,\frac{\mathrm{m}}{\mathrm{s}}$.

[6]Initially, ALMA was not designed for VLBI observations. In order to work as a phased array equivalent to a single big telescope, ALMA's processing unit that correlates the signals from the single antennas was changed.

Table 1.1 Radio telescopes in the VLBI network; CARMA was shut down in April, 2015; ALMA and SPT are not yet fully integrated, the GLT is under construction

Sign	Telescope	Location
PdBI	Plateau de Bure Interferometer	Grenoble, France
IRAM	Institut de Radioastronomie Millimetrique	Pico Veleta, Spain
SMT	Submillimeter Telescope Observatory	Mount Graham, Arizona
JCMT	James Clerk Maxwell Telescope	Mauna Kea, Hawaii
SMA	Submillimeter Array	Mauna Kea, Hawaii
LMT	Large Millimeter Telescope	Sierra Negra, Mexico
APEX	Atacama Pathfinder Experiment	Chajnantor-Plateau, Chile
CARMA	Combined Array for Research in Millimeter-wave Astronomy	Inyo Mountains, California
ALMA	Atacama Large Millimeter/submillimeter Array	Chajnantor-Plateau, Chile
SPT	South Pole Telescope	Amundsen–Scott Station, Antarctica
GLT	Greenland Telescope	Summit Station, Greenland

Table 1.2 Achievable angular resolutions δ for a maximal baseline of $D = 10000$ km for typical observing frequencies ν in the mm/sub-mm wavelength range

Frequency ν (GHz)	150	230	345	450	500
Wavelength λ (mm)	2	1.3	0.87	0.67	0.60
$\delta(10^4$ km) (μas)	41.22	26.89	17.92	13.74	12.37

One algorithm that is used is the CLEAN algorithm from Högbom (1974), see also the historic remarks and comments given by Högbom (2003) himself and Cornwell (2009).

Even though higher resolution is achieved with submillimeter wavelength astronomy, it is technically very difficult. In fact, observations are not disturbed by interstellar dust, but by atmospheric streams or by water that is contained in the atmosphere. For some wavelengths, the radiation (microwaves) is even completely absorbed, see for example Fig. 1 by Maiolino (2008) which shows the atmospheric transmission for the ALMA site in the Atacama Desert, Chile. This effect determines possible observing frequencies. Furthermore, the signals itself get Doppler shifted because of the Earth's rotation which complicates the comparison among the different locations.

Besides all technical challenges that have to be solved for imaging the shadow there are several other questions that have to be clarified first:

- How would the shadow look like?
- Will the contrast be high enough to see the shadow?
- How does an accretion disk or a jet affect the shadow?
- Is it possible to localize the targets with sufficient precision?

These are answered with the help of analytic and numerical calculations.[7]

Synge (1966) was the first to calculate what we nowadays call the shadow[8] of a Schwarzschild black hole. He found that the angular radius ρ of the circular shadow is given by the simple formula

$$\sin^2 \rho = \frac{27}{4} \frac{(\rho_O - 1)}{\rho_O^3} \tag{1.2}$$

where $\rho_O = r_O/(2\,\mathrm{m})$ is the ratio of the observer's r coordinate r_O and the Schwarzschild radius. Here, circular light orbits exist on the photon sphere at $r = 3\,\mathrm{m}$. For a rotating Kerr black hole, the photon sphere breaks into a spatially three-dimensional photon region filled by spherical lightlike geodesics, i.e., by light paths on spheres $r =$ constant. The shadow is non-circular and gets a D-shaped contour in the extreme case. Bardeen (1973) was the first to correctly calculate the shadow of a Kerr black hole; the results can also be found in Chandrasekhar's book or in Volker Perlick's Living Review (Chandrasekhar 1983; Perlick 2004). Bardeen's distant observer is suitable for describing the shape of the shadow. The size is not characterized.

The shadow has also been discussed for other black holes, e.g., for charged black holes (and for naked singularities) in the Kerr–Newman space-time (de Vries 2000), for $\delta = 2$ Tomimatsu–Sato space-times (Bambi and Yoshida 2010), for black holes in extended Chern–Simons modified gravity (Amarilla et al. 2010), in a Randall–Sundrum braneworld scenario (Amarilla and Eiroa 2012) and a Kaluza–Klein rotating dilaton black hole (Amarilla and Eiroa 2013), for the Kerr–NUT space-time (Abdujabbarov et al. 2012), for multi-black holes (Yumoto et al. 2012) and for regular black holes (Li and Bambi 2014). None of these descriptions is an analytic one.

Light rays near the central black hole are expected to be affected by the luminous accretion disk or other dust that surrounds the black hole, see Fig. 1.2 for a schematic illustration of the shadow. The matter causes a spreading such that the black hole's silhouette is not sharply delimited. In order to generate a more realistic image of the black hole's shadow, the visual appearance of an accretion disk or jets is studied with the help of sophisticated numerical ray-tracing programs by several authors, following the pioneering work of Bardeen and Cunningham (1973) and Luminet (1979). Here, opacity effects caused by absorption, emission or scattering could be modeled by general relativistic magnetohydrodynamics (GRMHD) or general relativistic radiative transfer (GRRT). Starting from a grid which represents the observed image, the resulting equations are solved backwards in time simultaneously with the geodesic equation.

There are numerous articles about these simulations. Since this doctoral thesis covers an analytic way to calculate the shadow neglecting all effects of matter, we refer only to a selection. Influences of jets, accretion disks/tori (optical density,

[7] The following two paragraphs are based on expositions in [1].

[8] Synge did not use the word "shadow" but he investigated the condition under which photons could escape to infinity; this complement of the shadow he called "escape cone".

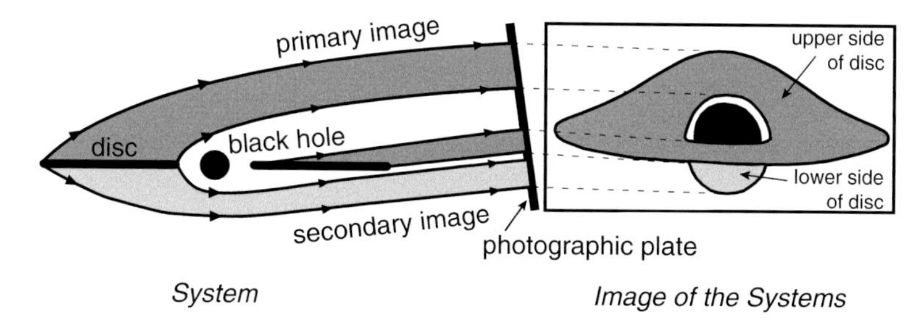

Fig. 1.2 Schematic illustration of the shadow of a *black hole* with accretion disk (*Credit* © J.-P. Luminet, original from Luminet 1998)

magnetic properties, polarization) or compton scattering at relativistic electrons were explored by Agol (1997), Armitage and Reynolds (2003), Broderick and Narayan (2006), Bromley et al. (2001), Dexter (2011), Dexter et al. (2012), Mościbrodzka et al. (2011, 2012, 2014), Vincent et al. (2015), Younsi et al. (2012), Younsi and Wu (2013). Furthermore, one can determine how the shadow looks like for an approaching observer (Marck 1996), during a star collapse (Ames and Thorne 1968; Ortiz et al. 2015) or for the collision or the merger of two black holes (Yumoto et al. 2012; Bohn et al. 2015). For the movie *Interstellar*, an algorithm for tracing ray-bundles instead of single photons was developed. It was used to generate high resolution sequences of the shadow of an almost maximally rotating Kerr black hole with an accretion disk (James et al. 2015). Finally, the paper by Falcke et al. (2000) has to be mentioned. Here, Heino Falcke, Fulvio Melia and Eric Agol demonstrate the effect of scattering on the visibility of the shadow. Since publication, this paper is quoted in virtually all publications regarding the black hole's shadow!

Besides the nearest candidate Sgr A* (8.3 kpc away with mass $4.3 \times 10^6 M_\odot$) another promising candidate for a supermassive black hole is the object at the center of M87 (16.7 Mpc away with mass $6.2 \times 10^9 M_\odot$), see Lu et al. (2014), Broderick et al. (2015) for black hole argument for M87 and Gillessen et al. (2009), Broderick et al. (2015), Kormendy and Ho (2013) for the black hole parameters. A broad overview of observations as well as simulations of phenomena for the black holes near Sgr A* and in M87 is given by Dexter and Fragile (2013).[1]

Using Eq. (1.2), we can estimate the angular diameter of shadows casted by the black holes in the Galactic center and in M87, see Table 4.1 in Sect. 4.6. According to that, one expects for Sgr A* and M87 angular diameters of 53 and 40 μas (microarcseconds) which are comparable to the angular diameter of an orange on the moon observed from the earth. Although tiny, such a diameter should be resolvable with VLBI; to achieve angular resolution of 20 μas, submillimeter wavelength observations at 0.87 mm (345 GHz) or even better at 0.67 mm are necessary, cf. Table 1.2. The submillimeter wavelength range has the advantage that these wavelengths are less distorted by interstellar scattering on galactic electrons (Ricarte and Dexter 2015). Just as important as the achievement of the required angular resolution is the question whether it is possible to localize SgrA* and M87 within some microarcseconds. This was answered positively by Broderick et al. (2011).

Fortunately, the numerical studies by Falcke et al. (2000) reveal that the shadow is visible at 0.6 mm submillimeter wavelength but not at 1.3 mm, see also Doeleman et al. (2008), Huang et al. (2007). In addition, there is a new paper by Fish et al. (2014). After generating a VLBI simulation at 1.3 mm wavelength by image processing tools, they show that image reconstruction by deconvolution is possible.

First VLBI observations at 1.3 mm (three-station VLBI: SMT, CARMA, JCMT) show structure in the order of the event horizon of the black hole near Sgr A* but did not resolve the shadow (Doeleman et al. 2008). The needed higher resolution is expected to be achieved in the near future when ALMA and SPT are fully integrated in the network. For sure, the first actually observed shadow will be posted on the internet blog *Dark Star Diaries*[9] established as public outreach from the EHT project; the already existing posts contain useful informations around the project.

In spite of the fact that first observations have not resolved the shadow, they do not disappoint. They do show that the bulk of the Sgr A* emission is slightly offset of the center of gravity. Consequently, the radio source Sgr A* does not coincide with the black hole. Doeleman et al. (2008) suggest that the emission arises in the surrounding accretion disk or jet. The observed offset is explained by doppler effects. Other observations confirm a small emission region (Krichbaum et al. 1998; Fish 2010), resolve a jet launching structure (Doeleman et al. 2012) or show that the accretion disk is tilted with aligned inclination (50°–60°) of source and the inner stellar disk (Dexter and Fragile 2013; Mościbrodzka et al. 2014; Psaltis et al. 2015).

Special interest has been devoted to the question of whether the shadow of a black hole can be used as a test of the *no-hair theorem*[1], which states that a black hole is completely characterized by mass, spin and electric charge (Misner et al. 1973). Recently, Norman Gürlebeck showed that the no-hair theorem is not only valid for isolated black holes but also for astrophysical situations where black holes are distorted by matter (Gürlebeck 2015; Ashtekar 2015). The observational aspects were mainly discussed by Tim Johannsen, see for example Johannsen and Psaltis (2011); Johannsen (2012a, b, 2013); Broderick et al. (2014). If the shadow of a black hole will be observed, its shape will give important information on the parameters of the black hole. There are considerations whether it is possible to constrain at least the spin from the shadow (Takahashi 2004; Zakharov et al. 2012; Tsukamoto et al. 2014). For this purpose, Hioki and Maeda (2009) introduced a deformation parameter that characterizes the deviation of the shadow from a circle.[1] Abdujabbarov et al. (2015) gave a more elaborated, coordinate independent characterization of the shadow.

Every few years, review articles about the current status and the future of VLBI appear; the reports of the last decade are focussed on the shadow observations (Spencer 1991; Moran 2003; Fish and Doeleman 2009; Doeleman 2009, 2010; Krichbaum et al. 2012). Observational perspectives for submm VLBI with ALMA are highlighted by Krichbaum (2010), Fish et al. (2013), Tilanus et al. (2014).

[9]Blog website: blogs.scientificamerican.com/dark-star-diaries/.

Since the beginning of the Event Horizon Telescope project it is aimed to extend the existing network of mm/submm telescopes in order to increase the resolution. This is done by deploying receivers for missing submm wavelengths in the simplest case, by upgrading existing telescopes, as for example ALMA, or by commissioning completely new VLBI stations. The most suitable sites are those with relatively stable atmospheric conditions, in particular with low fractions of water vapor. Thus, deserts or the arctic regions are highly favored. Indeed, after ALMA (in the Atacama Desert) and the SPT, the Greenland Telescope (GLT), which is currently under construction, will join the VLBI network (Nakamura et al. 2013; Inoue et al. 2014). Furthermore, there are plannings for a phase-a study for a new telescope in Africa (possible site: Namibia, Kili) within the BlackHoleCam project. If everything works well and the 10-meter space-based radio telescope *Millimetron* goes into operation at the Lagrangian point L2 probably in the mid-2020s (Kardashev et al. 2014), then the resolution will be further improved. With this Russian satellite, the Earth-based telescope network is upgraded with an extra-long Space-Earth baseline of 1.5 million km. Perhaps, it is then possible to observe shadows of other black holes candidates like those listed by Inoue et al. (2012).

For observing the position of stars, the breakthrough was succeeded with developing adaptive optics. With this technology, the atmospheric blurring effects are reduced notably by adjusting the (segmented) mirrors of the telescope depending on Laser measurements of the atmosphere. The observations of stellar orbits in the Galactic center are expected to become even more precise when the GRAVITY instrument at the Very Large Telescope (VLT) in Chile (Eisenhauer et al. 2007, 2009) goes into operation soon.[1] Higher resolutions are guaranteed with the next-generation optical/near-infrared telescope. Last year, the constructing phase of the new 40-meter European Extremely Large Telescope (EELT) started. It is also located in the Atacama Desert in Chile (Cerro Armazones) and *first light* is anticipated for the next decade (Lyubenova and Kissler-Patig 2011).

With these new telescopes, stellar orbits at 100 Schwarzschild radii could be discovered—more than one order of magnitude better than today. It will be also possible to study the origin of flares in more detail. After imaging the shadow of the black hole in the Galactic center near Sgr A* or that one in M87, the scientific main goal of the Event Horizon Telescope or the BlackHoleCam project is to estimate black hole parameter from the data like mass and distance or spin, orientation or a putative quadrupole moment for tests of general relativity.

Finally, Table 1.3 gives a chronological overview from the first considerations of gravitational light deflection by Cavendish and von Soldner to the subsequently presented analyitc description of the shadow of a black hole. Listed are important steps for the theoretical description and for the observational techniques needed to observe the shadow.

Table 1.3 Timetable

1784	Newtonian light deflection (H. Cavendish)	Will (1988)
1801	Newtonian light deflection (J. von Soldner)	von Soldner (1804)
1905	Special Relativity	Einstein (1905)
1915	General Relativity	Einstein (1915a, b, c)
1916	First solution of Einstein's field equations	Schwarzschild (1916)
1919	Confirmation of Einsteinian light deflection	Dyson et al. (1920)
1963	Discovery of the Kerr solution	Kerr (1963)
1964	Origin of the naming "black hole"	Ewing (1964)
1966	Shadow of a Schwarzschild black hole	Synge (1966)
1967	First VLBI observations	Kellermann (1972)
1971	Active galactic nuclei hosts supermassive black holes	Lynden-Bell and Rees (1971), Rees (1984)
1973	Shadow of a Kerr black hole	Bardeen (1973)
1973	"A black hole has no hair" (no-hair theorem)	Misner et al. (1973)
1974	Discovery of the radio source Sagittarius A* (Sgr A*) in the Galactic center	Balick and Brown (1974), Goss et al. (2003)
1979	First lensed double image of twin quasar QSO 0957+561	Walsh et al. (1979)
1979	Shadow of a black hole with accretion disk	Luminet (1979)
1992	First measurements of proper motion of stars in the Galactic center	Eckart and Genzel (1997), Ghez et al. (1998)
2000	Shadow of Kerr–Newman black hole and naked singularities	de Vries (2000)
2000	First ray-tracing simulation for Galactic black hole	Falcke et al. (2000)
2008	First NSF grant for the Event Horizon Telescope	
2013	ERC grant for the BlackHoleCam project	
2015	No-hair theorem for astrophysical black holes	Gürlebeck (2015)
2015	Shadows for moving observers, aberration	Grenzebach (2015)
2015	Shadows of black holes in the Plebański–Demiański class	Grenzebach et al. (2015)

By [1–3] I refer to my papers Grenzebach et al. (2014), Grenzebach (2015) and Grenzebach et al. (2015), respectively. Sentences marked with [i] can be found in total or only slightly modified in the ith paper

References

Abdujabbarov A, Atamurotov F, Kucukakca Y, Ahmedov B, Camci U (2012) Shadow of Kerr-Taub-NUT black hole. Astrophys Space Sci 344(2):429–435. doi:10.1007/s10509-012-1337-6

Abdujabbarov AA, Rezzolla L, Ahmedov BJ (2015) A coordinate-independent characterization of a black-hole shadow. Mon Not R Astron Soc 454(3):2423–2435. doi:10.1093/mnras/stv2079, arXiv:1503.09054

Agol E (1997) The effects of magnetic fields, absorption, and relativity on the polarization of accretion disks around supermassive black holes. Dissertation, University of California, Santa Barbara. http://faculty.washington.edu/agol/thesis.html

Amarilla L, Eiroa EF (2012) Shadow of a rotating braneworld black hole. Phys Rev D 85(6):064,019(9). doi:10.1103/PhysRevD.85.064019

Amarilla L, Eiroa EF (2013) Shadow of a Kaluza-Klein rotating dilaton black hole. Phys Rev D 87:044,057(7). doi:10.1103/PhysRevD.87.044057

Amarilla L, Eiroa EF, Giribet G (2010) Null geodesics and shadow of a rotating black hole in extended Chern-Simons modified gravity. Phys Rev D 81(12):124,045(8). doi:10.1103/PhysRevD.81.124045

Ames WL, Thorne KS (1968) The Optical appearance of a star that is collapsing through its gravitational radius. Astrophys J 151:659–670. doi:10.1086/149465

Armitage PJ, Reynolds CS (2003) The variability of accretion on to Schwarzschild black holes from turbulent magnetized discs. Mon Not R Astron Soc 341(3):1041–1050. doi:10.1046/j.1365-8711.2003.06491.x

Ashtekar A (2015) Viewpoint: the simplicity of black holes. Physics 8(34). doi:10.1103/Physics.8.34

Balick B, Brown RL (1974) Intense sub-arcsecond structure in the galactic center. Astrophys J 194:265–270. doi:10.1086/153242

Bambi C, Yoshida N (2010) Shape and position of the shadow in the $\delta = 2$ Tomimatsu–Sato spacetime. Class Quantum Gravity 27(20):205,006(10). doi:10.1088/0264-9381/27/20/205006

Bardeen JM (1973) Timelike and null geodesics in the Kerr metric. In: DeWitt C, DeWitt BS (eds) Black holes pp 215–239, New York

Bardeen JM, Cunningham CT (1973) The optical appearance of a star orbiting an extreme Kerr black hole. Astrophys J 183:237–264. doi:10.1086/152223

Bohn A, Hébert F, Throwe W, Bunandar D, Henriksson K, Scheel MA, Taylor NW (2015) What does a binary black hole merger look like? Class Quantum Gravity 32(6):065,002(16). doi:10.1088/0264-9381/32/6/065002, arXiv:1410.7775

Britzen S (2012) Verbotenes Universum: Die Zeit der Schwarzen Löcher. Goldegg Verlag, Wien

Broderick AE, Loeb A (2009) Portrait of a black hole. Sci Am 301:42–49. doi:10.1038/scientificamerican1209-42

Broderick AE, Narayan R (2006) On the Nature of the Compact Dark Mass at the Galactic Center. Astrophys J 638:L21–L24. doi:10.1086/500930

Broderick AE, Loeb A, Narayan R (2009) The event horizon of Sagittarius A*. Astrophys J 701(2):1357–1366. doi:10.1088/0004-637X/701/2/1357

Broderick AE, Loeb A, Reid MJ (2011) Localizing Sagittarius A* and M87 on Microarcsecond scales with millimeter VLBI. Astrophys J 735(1):57(18), doi:10.1088/0004-637X/735/1/57

Broderick AE, Johannsen T, Loeb A, Psaltis D (2014) Testing the no-hair theorem with event horizon telescope observations of Sagittarius A*. Astrophys J 784(1):7(14pp). doi:10.1088/0004-637X/784/1/7

Broderick AE, Narayan R, Kormendy J, Perlman ES, Rieke MJ, Doeleman SS (2015) The Event Horizon of M87. Astrophys J 805(2): doi:10.1088/0004-637X/805/2/179, arXiv:1503.03873

Bromley BC, Melia F, Liu S (2001) Polarimetric imaging of the massive black hole at the Galactic Center. Astrophys J 555(2):L83–L86. doi:10.1086/322862

Cardoso V, Crispino LCB, Macedo CFB, Okawa H, Pani P (2014) Light rings as observational evidence for event horizons: Long-lived modes, ergoregions and nonlinear instabilities of ultracompact objects. Phys Rev D 90:044,069(10). doi:10.1103/PhysRevD.90.044069

Chandrasekhar S (1983) The mathematical theory of black holes, International Series of Monographs on Physics, vol 69. Oxford University Press, Oxford

Cornwell TJ (2009) Hogbom's CLEAN algorithm. Impact on astronomy and beyond. Astron Astrophys 500(65–66):1974. doi:10.1051/0004-6361/200912148 (commentary on Högbom (1974))

Dexter J (2011) Radiative Models of Sagittarius A* and M87 from Relativistic MHD Simulations. Dissertation, University of Washington, Washington. http://hdl.handle.net/1773/17085

Dexter J, Fragile PC (2013) Tilted black hole accretion disc models of Sagittarius A*: time-variable millimetre to near-infrared emission. Mon Not R Astron Soc 432:2252–2272. doi:10.1093/mnras/stt583

Dexter J, Agol E, Fragile PC, McKinney JC (2012) Radiative models of Sagittarius A* and M87 from relativistic MHD simulations. J Phys: Conf Ser 372(1):012,023(8). doi:10.1088/1742-6596/372/1/012023

Doeleman S (2009) Imaging an Event Horizon: submm-VLBI of a Super Massive Black Hole. Astro2010: The Astronomy and Astrophysics Decadal Survey, Science White Papers 68, a Science White Paper to the Decadal Review Committee, arXiv:0906.3899

Doeleman S (2010) Building an Event Horizon Telescope: (sub)mm VLBI in the ALMA era. In: Proceedings of Science PoS (10th EVN Symposium)(053). http://pos.sissa.it/cgi-bin/reader/contribution.cgi?id=PoS(10th%20EVN%20Symposium)053 10th European VLBI Network Symposium and EVN Users Meeting: VLBI and the new generation of radio arrays, September 20 – 24, 2010, Manchester Uk

Doeleman SS, Weintroub J, Rogers AEE, Plambeck R, Freund R, Tilanus RPJ, Friberg P, Ziurys LM, Moran JM, Corey B, Young KH, Smythe DL, Titus M, Marrone DP, Cappallo RJ, Bock DCJ, Bower GC, Chamberlin R, Davis GR, Krichbaum TP, Lamb J, Maness H, Niell AE, Roy A, Strittmatter P, Werthimer D, Whitney AR, Woody D (2008) Event-horizon-scale structure in the supermassive black hole candidate at the Galactic Centre. Lett Nat 455:78–80. doi:10.1038/nature07245

Doeleman SS, Fish VL, Schenck DE, Beaudoin C, Blundell R, Bower GC, Broderick AE, Chamberlin R, Freund R, Friberg P, Gurwell MA, Ho PTP, Honma M, Inoue M, Krichbaum TP, Lamb J, Loeb A, Lonsdale C, Marrone DP, Moran JM, Oyama T, Plambeck R, Primiani RA, Rogers AEE, Smythe DL, SooHoo J, Strittmatter P, Tilanus RPJ, Titus M, Weintroub J, Wright M, Young KH, Ziurys LM (2012) Jet-launching structure resolved near the supermassive black hole in M87. Science 338(6105):355–358. doi:10.1126/science.1224768

Dyson FW, Eddington AS, Davidson C (1920) A determination of the deflection of light by the sun's gravitational field, from observations made at the total eclipse of May 29, 1919. Philos Trans R Soc Lond A: Math Phys Eng Sci 220(571–581):291–333. doi:10.1098/rsta.1920.0009

Eckart A, Genzel R (1996) Observations of stellar proper motions near the Galactic Centre. Nature 383(5):415–417. doi:10.1038/383415a0

Eckart A, Genzel R (1997) Stellar proper motions in the central 0.1 pc of the Galaxy. Mon Not R Astron Soc 284(3):576–598. doi:10.1093/mnras/284.3.576

Eckart A, Schödel R, Straubmeier C (2005) The black hole at the center of the Milky Way. Imperial College Press, Covent Garden. doi:10.1142/9781860947391

Einstein A (1905) Zur Elektrodynamik bewegter Körper. Ann der Phys 17:891–921. http://www.physik.uni-augsburg.de/annalen/history/einstein-papers/1905_17_891-921.pdf

Einstein A (1915a) Die Feldgleichungen der Gravitation. Sitzungsberichte der Königlich-Preußischen Akademie der Wissenschaften, pp 844–847. http://echo.mpiwg-berlin.mpg.de/MPIWG:ZZB2HK6W

Einstein A (1915b) Zur allgemeinen Relativitätstheorie. Sitzungsberichte der Königlich-Preußischen Akademie der Wissenschaften, pp 778–786. http://echo.mpiwg-berlin.mpg.de/MPIWG:DPP4MDQV

Einstein A (1915c) Zur allgemeinen Relativitätstheorie (Nachtrag). Sitzungsberichte der Königlich-Preußischen Akademie der Wissenschaften, pp 799–801. http://echo.mpiwg-berlin.mpg.de/MPIWG:GTN3GYS8

Einstein A (1916) Die Grundlage der allgemeinen Relativitätstheorie. Ann der Phys 49(7):769–822. http://www.physik.uni-augsburg.de/annalen/history/einstein-papers/1916_49_769-822

Eisenhauer F, Schödel R, Genzel R, Ott T, Tecza M, Abuter R, Eckart A, Alexander T (2003) A Geometric Determination of the Distance to the Galactic Center. Astrophys J Lett 597(2):L121–L124. doi:10.1086/380188

Eisenhauer F, Perrin G, Straubmeier C, Brandner W, Boehm A, Cassaing F, Clenet Y, Dodds-Eden K, Eckart A, Fedou P, Gendron E, Genzel R, Gillessen S, Graeter A, Gueriau C, Hamaus N, Haubois X, Haug M, Henning T, Hippler S, Hofmann R, Hormuth F, Houairi K, Kellner S, Kervella P, Klein R, Kolmeder J, Laun W, Lena P, Lenzen R, Marteaud M, Meschke D, Naranjo V, Neumann U, Paumard T, Perger M, Perret D, Rabien S, Ramos JR, Reess JM, Rohloff RR, Rouan D, Rousset G, Ruyet B, Schropp M, Talureau B, Thiel M, Ziegleder J, Ziegler D (2007) GRAVITY: microarcsecond astrometry and deep interferometric imaging with the VLTI. In: Proceedings of the International Astronomical Union, symposium No. 248, vol 3, pp 100–101. doi:10.1017/S1743921308018723

Eisenhauer F, Perrin G, Brandner W, Straubmeier C, Böhm A, Baumeister H, Cassaing F, Clénet Y, Dodds-Eden K, Eckart A, Gendron E, Genzel R, Gillessen S, Gräter A, Gueriau C, Hamaus N, Haubois X, Haug M, Henning T, Hippler S, Hofmann R, Hormuth F, Houairi K, Kellner S, Kervella P, Klein R, Kolmeder J, Laun W, Léna P, Lenzen R, Marteaud M, Naranjo V, Neumann U, Paumard T, Rabien S, Ramos JR, Reess JM, Rohloff D R-R Rouan, Rousset G, Ruyet B, Sevin A, Thiel M, Ziegleder J, Ziegler D (2009) GRAVITY: Microarcsecond Astrometry and Deep Interferometric Imaging with the VLT. In: Moorwood (2009), pp 361–365. doi:10.1007/978-1-4020-9190-2_61

Ewing A (1964) 'Black Holes' in space. Sci News Lett 85(3):39. https://www.sciencenews.org/archive/black-holes-space

Falcke H, Melia F, Agol E (2000) Viewing the shadow of the black hole at the Galactic Center. Astrophys J 528:L13–L16. doi:10.1086/312423

Falcke H, Hehl F (eds) (2003) The galactic black hole: lectures on general relativity and astrophysics. Cosmology and Gravitation, IoP Publishing, Bristol, U.K, Series in High Energy Physics

Falcke H, Markoff SB (2013) Toward the event horizon—the supermassive black hole in the Galactic Center. Class Quantum Gravity 30(24):244,003(24pp). doi:10.1088/0264-9381/30/24/244003, published in "Astrophysical black holes", ed. by D. Merritt and L. Rezzolla

Fish V (2010) Observing event horizons with high-frequency VLBI. In: Proceedings of Science PoS (10th EVN Symposium). 10th european VLBI network symposium and EVN users meeting: VLBI and the new generation of radio arrays, 20–24 Sept 2010, Manchester, Uk. http://pos.sissa.it/cgi-bin/reader/contribution.cgi?id=PoS(10th%20EVN%20Symposium)052

Fish VL, Doeleman SS (2009) Observing a black hole event horizon: (sub)millimeter VLBI of Sgr A*. In: Proceedings of the International Astronomical Union, vol 5, pp 271–276. doi:10.1017/S1743921309990500, symposium S261 (Relativity in Fundamental Astronomy: Dynamics, Reference Frames, and Data Analysis)

Fish V, Alef W, Anderson J, Asada K, Baudry A, Broderick A, Carilli C, Colomer F, Conway J, Dexter J, Doeleman S, Eatough R, Falcke H, Frey S, Gabányi K, Gálvan-Madrid R, Gammie C, Giroletti M, Goddi C, Gómez JL, Hada K, Hecht M, Honma M, Humphreys E, Impellizzeri V, Johannsen T, Jorstad S, Kino M, Körding E, Kramer M, Krichbaum T, Kudryavtseva N, Laing R, Lazio J, Loeb A, Lu RS, Maccarone T, Marscher A, Martí-Vidal I, Martins C, Matthews L, Menten K, Miller J, Miller-Jones J, Mirabel F, Muller S, Nagai H, Nagar N, Nakamura M, Paragi Z, Pradel N, Psaltis D, Ransom S, Rodríguez L, Rottmann H, Rushton A, Shen ZQ, Smith D, Stappers B, Takahashi R, Tarchi A, Tilanus R, Verbiest J, Vlemmings W, Walker RC, Wardle J, Wiik K, Zackrisson E, Zensus JA (2013) High-Angular-Resolution and High-Sensitivity Science Enabled by Beamformed ALMA, whitepaper, arXiv:1309.3519

Fish VL, Johnson MD, Lu RS, Doeleman SS, Bouman KL, Zoran D, Freeman WT, Psaltis D, Narayan R, Pankratius V, Broderick AE, Gwinn CR, Vertatschitsch LE (2014) Imaging an Event Horizon: mitigation of scattering toward Sagittarius A*. Astrophys J 795:134(7pp). doi:10.1088/0004-637X/795/2/134

Genzel R (2014) Massive black holes: evidence, demographics and cosmic evolution, In: Blandford R, Sevrin A (eds) Proceedings of the 26th solvay conference on physics: "Astrophysics and Cosmology". World Scientific, arXiv:1410.8717

Genzel R, Eisenhauer F, Gillessen S (2010) The Galactic Center massive black hole and nuclear star cluster. Rev Mod Phys 82:3121. doi:10.1103/RevModPhys.82.3121, arXiv:1006.0064

Ghez AM, Klein BL, Morris M, Becklin EE (1998) High Proper-Motion Stars in the Vicinity of Sagittarius A*: Evidence for a Supermassive Black Hole at the Center of Our Galaxy. Astrophys J 509(2):678. doi:10.1086/306528

Ghez AM, Salim S, Hornstein SD, Tanner A, Lu JR, Morris M, Becklin EE, Duchêne G (2005) Stellar Orbits around the Galactic Center Black Hole. Astrophys J 620(2):744. doi:10.1086/427175

Gillessen S, Eisenhauer F, Quataert E, Genzel R, Paumard T, Trippe S, Ott T, Abuter R, Eckart A, Lagage PO, Lehnert MD, Tacconi LJ, Martins F (2006) Variations in the spectral slope of Sagittarius A* during a Near-Infrared Flare. Astrophys J Lett 640(2):L163. doi:10.1086/503557

Ghez AM, Salim S, Weinberg NN, Lu JR, Do T, Dunn JK, Matthews K, Morris MR, Yelda S, Becklin EE, Kremenek T, Milosavljevic M, Naiman J (2008) Measuring Distance and Properties of the Milky Way's Central Supermassive Black Hole with Stellar Orbits. Astrophys J 689(2):1044–1062. doi:10.1086/592738

Gillessen S, Eisenhauer F, Trippe S, Alexander T, Genzel R, Martins F, Ott T (2009) Monitoring Stellar Orbits around the Massive Black Hole in the Galactic Center. Astrophys J 692(2):1075–1109. doi:10.1088/0004-637X/692/2/1075

Goss WM, Brown RL, Lo KY (2003) The Discovery of Sgr A*. Astron Nachr 324(S1):497–504. doi:10.1002/asna.200385047 (Proceedings of the Galactic Center Workshop 2002—The central 300 parsecs of the Milky Way, arXiv:astro-ph/0305074)

Grenzebach A, Perlick V, Lämmerzahl C (2014) Photon regions and shadows of Kerr–Newman–NUT Black Holes with a cosmological constant. Phys Rev D 89:124,004(12). doi:10.1103/PhysRevD.89.124004. arXiv:1403.5234

Grenzebach A (2015) Aberrational effects for shadows of black holes. In: Puetzfeld et al Proceedings of the 524th WE-Heraeus-Seminar "Equations of Motion in Relativistic Gravity", held in Bad Honnef, Germany, 17–23 Feb 2013, pp 823–832. doi:10.1007/978-3-319-18335-0_25, arXiv:1502.02861

Grenzebach A, Perlick V, Lämmerzahl C (2015) Photon regions and shadows of accelerated black holes. Int J Mod Phys D 24(9):1542,024(22). doi:10.1142/S0218271815420249 ("Special Issue Papers" of the "7th Black Holes Workshop", Aveiro, Portugal, arXiv:1503.03036)

Gürlebeck N (2015) No-Hair Theorem for Black Holes in Astrophysical Environments. Phys Rev Lett 114:151,102(5). doi:10.1103/PhysRevLett.114.151102, arXiv:1503.03240

Hawking SW (1974) Black hole explosions? Nature 248:30–31. doi:10.1038/248030a0

Hawking SW (1975) Particle creation by black holes. Commun Math Phys 43(3):199–220. doi:10.1007/BF02345020

Hawking SW (1976) Black holes and thermodynamics. Phys Rev D 13(2):191–197. doi:10.1103/PhysRevD.13.191

Hioki K, Maeda Ki (2009) Measurement of the Kerr spin parameter by observation of a compact object's shadow. Phys Rev D 80(2):024,042(9). doi:10.1103/PhysRevD.80.024042

Högbom JA (1974) Aperture synthesis with a non-regular distribution of interferometer baselines. Astron Astrophys Suppl 15:417–426

Högbom JA (2003) Early Work in Imaging. In: Zensus et al. Proceedings of a conference in honor of Kenneth I. Kellermann on the occasion of his 65th Birthday held at the National Radio Astronomy Observatory, Green Bank, West Virginia, USA, 10–12 Oct 2002

Huang L, Cai M, Shen ZQ, Yuan F (2007) Black hole shadow image and visibility analysis of Sagittarius A^*. Mon Not R Astron Soc 379(3):833–840. doi:10.1111/j.1365-2966.2007.11713.x

Inoue M, Blundell R, Brisken W, Chen MT, Doeleman S, Fish V, Ho P, Moran J, Napier P, the Greenland Telescope (GLT) Team (2012) Submm VLBI toward Shadow Image of Super Massive Black Hole. In: Proceedings of Science RTS2012(018). http://pos.sissa.it/cgi-bin/reader/contribution.cgi?id=PoS(RTS2012)018, resolving the Sky—Radio Interferometry: Past, Present and Future, 17–20 April 2012, Manchester Uk

Inoue M, Algaba-Marcos JC, Asada K, Blundell R, Brisken W, Burgos R, Chang CC, Chen MT, Doeleman SS, Fish V, Grimes P, Han J, Hirashita H, Ho PTP, Hsieh SN, Huang T, Jiang H, Keto E, Koch PM, Kubo DY, Kuo CY, Liu B, Martin-Cocher P, Matsushita S, Meyer-Zhao Z, Nakamura M, Napier P, Nishioka H, Nystrom G, Paine S, Patel N, Pradel N, Pu HY, Raffin PA, Shen HY, Snow W, Srinivasan R, Wei TS (2014) Greenland telescope project: direct confirmation of black hole with sub-millimeter VLBI. Radio Sci 49(7):564–571. doi:10.1002/2014RS005450, arXiv:1407.2450

James O, von Tunzelmann E, Paul F, Thorne KS (2015) Gravitational lensing by spinning black holes in astrophysics, and in the movie Interstellar. Class Quantum Gravity 32(6):065,001(41). doi:10.1088/0264-9381/32/6/065001

Johannsen T (2012a) Testing General Relativity in the Strong-Field Regime with Observations of Black Holes in the Electromagnetic Spectrum. Phd thesis, University of Arizona, Tucson. http://hdl.handle.net/10150/238893

Johannsen T (2012b) Testing the no-hair theorem with Sgr A*. Adv Astron 486750(9). doi:10.1155/2012/486750

Johannsen T (2013) Photon Rings around Kerr and Kerr-like black holes. Astrophys J 777(2):170(12). doi:10.1088/0004-637X/777/2/170

Johannsen T, Psaltis D (2011) Metric for rapidly spinning black holes suitable for strong-field tests of the no-hair theorem. Phys Rev D 83(12):124,015(10). doi:10.1103/PhysRevD.83.124015

Kardashev NS, Novikov ID, Lukash VN, Pilipenko SV, Mikheeva EV, Bisikalo DV, Wiebe DS, Doroshkevich AG, Zasov AV, Zinchenko II, Ivanov PB, Kostenko VI, Larchenkova TI, Likhachev SF, Malov IF, Malofeev VM, Pozanenko AS, Smirnov AV, Sobolev AM, Cherepashchuk AM, Shchekinov YA (2014) Review of scientific topics for Millimetron space observatory. Uspekhi Fizicheskih Nauk 184(12):1319–1352. doi:10.3367/UFNr.0184.201412c.1319, english translation available on, arXiv:1502.06071

Kellermann KI (1972) Intercontinental radio astronomy. Sci Am 226:72–83. doi:10.1038/scientificamerican0272-72

Kennefick D (2009) Testing relativity from the 1919 eclipse—a question of bias. Phys Today 37–42. doi:10.1063/1.3099578

Kerr RP (1963) Gravitational field of a spinning mass as an example of algebraically special metrics. Phys Rev Lett 11(5):237–238. doi:10.1103/PhysRevLett.11.237

Kormendy J, Ho LC (2013) Coevolution (Or Not) of Supermassive black holes and host galaxies. Ann Rev Astron Astrophys 51:511–653. doi:10.1146/annurev-astro-082708-101811, arXiv:1304.7762

Krichbaum TP, Graham DA, Witzel A, Greve A, Wink JE, Grewing M, Colomer F, de Vicente P, Gómez-González J, Baudry A, Zensus JA (1998) VLBI observations of the galactic center source Sgr A* at 86 GHz and 215 GHz. Astron Astrophys 335(3):L106–L110. http://aa.springer.de/bibs/8335003/230l106/small.htm

Krichbaum TP (2010) Imaging Super Massive Black Holes and the Origin of Jets—Global mm- and sub-mm-VLBI Studies of Compact Radio Sources. http://www.mpifr-bonn.mpg.de/div/vlbi/globalmm/pspdf/1mmwhitepaper.APEX-ALMA.pdf, a Whitepaper and Proposal for submm-VLBI with APEX and ALMA, Max-Planck-Institut für Radioastronomie, Bonn, Germany

Krichbaum TP, Roy A, Wagner J, Rottmann H, Hodgson JA, Bertarini A, Alef W, Zensus JA, Marscher A, Jorstad S, Freund R, Marrone D, Strittmatter P, Ziurys L, Blundell R, Weintroub J, Young K, Fish V, Doeleman S, Bremer M, Sanchez S, Fuhrmann L, Angelakis E, Karamanavis V (2012) Zooming towards the Event Horizon—mm-VLBI today and tomorrow.

Proc Sci 178(055). http://pos.sissa.it/cgi-bin/reader/contribution.cgi?id=PoS(11th%20EVN%20Symposium)055 Proceedings of the 11th European VLBI Network Symposium & Users Meeting, 20–24 Sept, Bordeaux, France
Kruesi L (2012) How we know black holes exist. Astronomy Magazine April:24–29. http://www.astronomy.com/-/media/Files/PDF/Magazine%20articles/How-we-know-black-holes-exist.pdf
Li Z, Bambi C (2014) Measuring the Kerr spin parameter of regular black holes from their shadow. J Cosmol Astropart Phys 01(041). doi:10.1088/1475-7516/2014/01/041
Lu RS, Broderick AE, Baron F, Monnier JD, Fish VL, Doeleman SS, Pankratius V (2014) Imaging the supermassive black hole shadow and jet base of M87 with the event horizon telescope. Astrophys J 788(2):120(10pp). doi:10.1088/0004-637X/788/2/120
Luminet JP (1979) Image of a spherical black hole with thin accretion disk. Astron Astrophys 75(1-2):228–235. http://adsabs.harvard.edu/abs/1979A%26A....75..228L
Luminet JP (1998) Black Holes: A General Introduction. In: Hehl et al Proceedings of the 179th W.E. Heraeus Seminar held at Bad Honnef, Germany, 18–22 August 1997. pp 3–34. doi:10.1007/978-3-540-49535-2_1, arXiv:astro-ph/9801252
Lynden-Bell D (1969) Galactic nuclei as collapsed old quasars. Nature 223:690–694. doi:10.1038/223690a0
Lynden-Bell D, Rees MJ (1971) On Quasars, Dust and the Galactic Centre. Mon Not R Astron Soc 152(4):461–475. doi:10.1093/mnras/152.4.461
Lyubenova M, Kissler-Patig M (eds) (2011) An Expanded View of the Universe—Science with the European Extremely Large Telescope. European Southern Observatory—E-ELT Science Office, Garching. http://www.eso.org/sci/facilities/eelt/science/doc/eelt_sciencecase.pdf
Maiolino R (2008) Prospects for AGN studies with ALMA. New Astron Rev 52:339–357. doi:10.1016/j.newar.2008.06.012
Marck JA (1996) Short-cut method of solution of geodesic equations for Schwarzschild black hole. Class Quantum Gravity 13:393–402. doi:10.1088/0264-9381/13/3/007
Mazur PO, Mottola E (2001) Gravitational Condensate stars: an alternative to black holes, unpublished, arXiv:gr-qc/0109035
Melia F (2003) The black hole at the center of our galaxy. Princeton University Press, Princeton. http://press.princeton.edu/titles/7480.html
Melia F, Falcke H (2001) The supermassive black hole at the Galactic Center. Ann Rev Astron Astrophys 39:309–352. doi:10.1146/annurev.astro.39.1.309
Meyer L, Ghez AM, Schödel R, Yelda S, Boehle A, Lu JR, Do T, Morris MR, Becklin EE, Matthews K (2012a) The Shortest-Known-Period Star Orbiting Our Galaxy's Supermassive Black Hole. Science 338(6103):84–87. doi:10.1126/science.1225506
Misner CW, Thorne KS, Wheeler JA (1973) Gravitation. W. H, Freeman and Company, San Francisco
Moran JM (2003) Thiry years of VLBI: early days, successes, and future. In: Zensus et al proceedings of a conference in honor of Kenneth I. Kellermann on the occasion of his 65th Birthday held at the National Radio Astronomy Observatory, pp 1–10, Green Bank, West Virginia, USA, 10–12 Oct 2002
Morris MR, Meyer L, Ghez AM (2012) Galactic center research: manifestations of the central black hole. Res Astron Astrophys 12(8):995–1020. doi:10.1088/1674-4527/12/8/007
Mościbrodzka M, Gammie CF, Dolence JC, Shiokawa H (2011) Pair production in low-luminosity galactic nuclei. Astrophys J 735(1):9(14). doi:10.1088/0004-637X/735/1/9
Mościbrodzka M, Shiokawa H, Gammie CF, Dolence JC (2012) The Galactic Center weather forecast. Astrophys J Lett 752(1):L1(6). doi:10.1088/2041-8205/752/1/L1
Mościbrodzka M, Falcke H, Shiokawa H, Gammie CF (2014) Observational appearance of inefficient accretion flows and jets in 3D GRMHD simulations: application to Sgr A*. Astron Astrophys 570:A7(10). doi:10.1051/0004-6361/201424358

Müller A (2004) Black Hole Astrophysics. Magnetohydrodynamics on the Kerr Geometry. Dissertation, University of Heidelberg, Heidelberg. http://www.wissenschaft-online.de/astrowissen/downloads/PhD/PhD_AMueller.pdf

Nakamura M, Algaba JC, Asada K, Chen B, Chen MT, Han J, Ho PHP, Hsieh SN, Huang T, Inoue M, Koch P, Kuo CY, Martin-Cocher P, Matsushita S, Meyer-Zhao Z, Nishioka H, Nystrom G, Pradel N, Pu HY, Raffin P, Shen HY, Tseng CY, the Greenland Telescope Project Team (2013) Greenland telescope project: a direct confirmation of black hole with submillimeter VLBI. EPJ Web Conf 61(01):008. doi:10.1051/epjconf/20136101008, arXiv:1310.1665

Ortiz N, Sarbach O, Zannias T (2015) The shadow of a naked singularity. Phys Rev D 92(044):035. doi:10.1103/PhysRevD.92.044035, arXiv:1505.07017

Petri M (2003a) Compact anisotropic stars with membrane—a new class of exact solutions to the Einstein field equations, unpublished, arXiv:gr-qc/0306063

Petri M (2003b) The holostar—a self-consistent model for a compact self-gravitating object, unpublished, arXiv:gr-qc/0306066

Perlick V (2004) Gravitational lensing form a spacetime perspective. Living Rev Relativ 7(9). doi:10.12942/lrr-2004-9

Petit JP (1995) Das Schwarze Loch. Die Abenteurer des Anselm Wüßtegern, Vieweg, Braunschweig; Wiesbaden. http://www.savoir-sans-frontieres.com/JPP/telechargeables/free_downloads.htm

Psaltis D, Narayan R, Fish VL, Broderick AE, Loeb A, Doeleman SS (2015) Event Horizon telescope evidence for alignment of the black hole in the center of the Milky Way with the inner stellar disk. Astrophys J 798(1):15. doi:10.1088/0004-637X/798/1/15

Rees MJ (1974) Black holes. Observatory 94:168–179

Rees MJ (1984) Black Hole Models for Active Galactic Nuclei. Ann Rev Astron Astrophys 22(1):471–506. doi:10.1146/annurev.aa.22.090184.002351

Ricarte A, Dexter J (2015) The Event Horizon Telescope: exploring strong gravity and accretion physics. Mon Not R Astron Soc 446(2):1973–1987. doi:10.1093/mnras/stu2128, arXiv:1410.2899

Salim S, Gould A (1999) Sagittarius A* "Visual Binaries": a direct measurement of the galactocentric distance. Astrophys J 523(2):633–641. doi:10.1086/307756

Schödel R, Eckart A, Straubmeier C, Pott JU (2005) NIR Observations of the Galactic Center. In: Röser S (ed) From cosmological structures to the Milky Way, Reviews in Modern Astronomy, vol 18. Wiley-VCH, Weinheim, pp 195–203. http://www.astronomische-gesellschaft.de/de/publikationen/reviews/reviews-in-modern-astronomy-18/Schoedel.pdf

Schwarzschild K (1916) Über das Gravitationsfeld eines Massenpunktes nach der Einsteinschen Theorie. Sitzungsberichte der Preussischen Akademie der Wissenschaften zu Berlin (VII):189–196. http://de.wikisource.org/wiki/Index:K._Schwarzschild_-_%C3%9Cber_das_Gravitationsfeld_eines_Massenpunktes_nach_der_Einsteinschen_Theorie_%281916%29.pdf digitale Volltext-Ausgabe in Wikisource

von Soldner JG (1804) Ueber die Ablenkung eines Lichtstrals von seiner geradlinigen Bewegung. Astronomisches Jahrbuch für das Jahr 1804, pp 161–172. https://de.wikisource.org/wiki/Ueber_die_Ablenkung_eines_Lichtstrals_von_seiner_geradlinigen_Bewegung

Spencer RE (1991) Very Long Baseline Interferometry: current status and future prospects. Vistas Astron 34:61–68. doi:10.1016/0083-6656(91)90020-S

Synge JL (1966) The escape of photons from gravitationally intense stars. Monthly Not R Astron Soc 131:463–466. http://dx.doi.org/10.1093/mnras/131.3.463

Takahashi R (2004) Shapes and Positions of Black Hole Shadows in Accretion Disks and Spin Parameters of Black Holes. Astrophys J 611(2):996–1004. doi:10.1086/422403

Tanaka Y, Nandra K, Fabian AC, Inoue H, Otani C, Dotani T, Hayashida K, Iwasawa K, Kii T, Kunieda H, Makino F, Matsuoka M (1995) Gravitationally redshifted emission implying an accretion disk and massive black hole in the active galaxy MCG-6-30-15. Lett Nat 375:659–661. doi:10.1038/375659a0

Thiébaut E (2009) Image reconstruction with optical interferometers. New Astron Rev 53(11–12):312–328. doi:10.1016/j.newar.2010.07.011 (proceedings: VLTI summerschool)

Thompson AR, Moran JM, Swenson GW Jr (2004) Interferometry and synthesis in radio astronomy, 2nd edn. Wiley, Weinheim. doi:10.1002/9783527617845

Tilanus RPJ, Krichbaum TP, Zensus JA, Baudry A, Bremer M, Falcke H, Giovannini G, Laing R, van Langevelde HJ, Vlemmings W (2014) Future mmVLBI Research with ALMA: A European vision, whitepaper on mm-VLBI with ALMA, arXiv:1406.4650

Tsukamoto N, Li Z, Bambi C (2014) Constraining the spin and the deformation parameters from the black hole shadow. J Cosmol Astropart Phys 6:043(21). doi:10.1088/1475-7516/2014/06/043, arXiv:1403.0371

Vincent FH, Yan W, Straub O, Zdziarski AA, Abramowicz MA (2015) A magnetized torus for modeling Sagittarius A* millimeter images and spectra. Astron Astrophys 574(A48): doi:10.1051/0004-6361/201424306

de Vries A (2000) The apparent shape of a rotating charged black hole, closed photon orbits and the bifurcation set A_4. Class Quantum Gravity 17:123–144. doi:10.1088/0264-9381/17/1/309

Walsh D, Carswell RF, Weymann RJ (1979) 0957 + 561 A, B: twin quasistellar objects or gravitational lens? Nature 279:381–384. doi:10.1038/279381a0

Will CM (1988) Henry Cavendish, Johann von Soldner, and the deflection of light. Am J Phys 56(5):413–415. doi:10.1119/1.15622

Younsi Z, Wu K (2013) Covariant Compton scattering kernel in general relativistic radiative transfer. Mon Not R Astron Soc 433(2):1054–1081. doi:10.1093/mnras/stt786

Younsi Z, Wu K, Fuerst SV (2012) General relativistic radiative transfer: formulation and emission from structured tori around black holes. Astron Astrophys 545:A13(3). doi:10.1051/0004-6361/201219599

Yumoto A, Nitta D, Chiba T, Sugiyama N (2012) Shadows of multi-black holes: analytic exploration. Phys Rev D 86(10):103,001(10). doi:10.1103/PhysRevD.86.103001

Zakharov AF, De Paolis F, Ingrosso G, Nucita AA (2012) Shadows as a tool to evaluate black hole parameters and a dimension of spacetime. New Astron Rev 56(2–3):64–73. doi:10.1016/j.newar.2011.09.002

Chapter 2
The Plebański–Demiański Class of Black Hole Space-Times

Abstract The Plebański–Demiański class contains stationary, axially symmetric type D solutions of the Einstein–Maxwell equations with a cosmological constant. It covers many well-known black hole space-times like the Schwarzschild, Kerr or the Kottler space-time. The space-times are characterized by seven parameters: mass, spin, electric and magnetic charge, gravitomagnetic NUT charge, a so-called acceleration parameter and the cosmological constant. We review space-time properties like symmetries and isometries as well as the appearance of singularities as ring singularities or axial singularities. Furthermore, we discuss horizons, the ergoregion and a region with causality violation.

Keywords Plebanski-Demianski · Schwarzschild · Kerr · Kerr-Newman · Reissner-Nordstroem · NUT · C-metric · Metric tensor · Boyer-Lindquist coordinates · Space-time properties · Symmetries · Isometries · Singularities · Ring singularity · Axial singularity · Black hole horizon · Ergoregion · Causality violation · Conformal factor

We consider the general Plebański–Demiański class of stationary, axially symmetric type D solutions of the Einstein–Maxwell equations with a cosmological constant.[1] In fact, these solutions were first found by Debever (1971) but are better known in the form of Plebański and Demiański (1976). For the case without cosmological constant, these metrics can be traced back to Carter (1968) and in the Boyer–Lindquist coordinates, which we will use in the following, to Miller (1973). A detailed discussion of the Plebański–Demiański metrics can be found in the books by Griffiths and Podolský (2009) or Stephani et al. (2003). It is common to use units in which the speed of light and Newtons gravitational constant are normalized ($c = 1$, $G = 1$). With this rescaling, the Plebański–Demiański metric can be written in the Boyer–Lindquist coordinates $(r, \vartheta, \varphi, t)$ as

[1] The first four paragraphs as well as parts of Sect. 2.3 are based on my papers [1] and [3]. Section 2.4 contains expositions given in [3].

A. Grenzebach, *The Shadow of Black Holes*,
SpringerBriefs in Physics, DOI 10.1007/978-3-319-30066-5_2

$$g_{\mu\nu}\,\mathrm{d}x^\mu\,\mathrm{d}x^\nu = \frac{1}{\Omega^2}\Big(\Sigma\big(\tfrac{1}{\Delta_r}\,\mathrm{d}r^2 + \tfrac{1}{\Delta_\vartheta}\,\mathrm{d}\vartheta^2\big) + \frac{1}{\Sigma}\Big((\Sigma + a\chi)^2\Delta_\vartheta\sin^2\vartheta - \Delta_r\chi^2\Big)\,\mathrm{d}\varphi^2 + \frac{2}{\Sigma}\Big(\Delta_r\chi - a(\Sigma + a\chi)\Delta_\vartheta\sin^2\vartheta\Big)\,\mathrm{d}t\,\mathrm{d}\varphi - \frac{1}{\Sigma}\Big(\Delta_r - a^2\Delta_\vartheta\sin^2\vartheta\Big)\,\mathrm{d}t^2\Big), \tag{2.1}$$

see Griffiths and Podolský (2009, p. 311). Here, we use the abbreviations

$$\Omega = 1 - \tfrac{\alpha}{\omega}(\ell + a\cos\vartheta)r, \qquad \Sigma = r^2 + (\ell + a\cos\vartheta)^2, \qquad \chi = a\sin^2\vartheta - 2\ell(\cos\vartheta + C),$$
$$\Delta_\vartheta = 1 - a_3\cos\vartheta - a_4\cos^2\vartheta, \qquad \Delta_r = b_0 + b_1 r + b_2 r^2 + b_3 r^3 + b_4 r^4 \tag{2.2}$$

with the following coefficients of the polynomials Δ_ϑ and Δ_r

$$a_3 = 2am\tfrac{\alpha}{\omega} - 4a\ell\big(\tfrac{\alpha^2}{\omega^2}(k+\beta) + \tfrac{\Lambda}{3}\big), \qquad a_4 = -a^2\big(\tfrac{\alpha^2}{\omega^2}(k+\beta) + \tfrac{\Lambda}{3}\big), \tag{2.3}$$

$$\begin{aligned} b_0 &= k + \beta, \\ b_1 &= -2m, \\ b_2 &= \tfrac{k}{a^2-\ell^2} + 4\tfrac{\alpha}{\omega}\ell m - (a^2 + 3\ell^2)\big(\tfrac{\alpha^2}{\omega^2}(k+\beta) + \tfrac{\Lambda}{3}\big), \\ b_3 &= -2\tfrac{\alpha}{\omega}\Big(\tfrac{k\ell}{a^2-\ell^2} - (a^2-\ell^2)\Big(\tfrac{\alpha}{\omega}m - \ell\big(\tfrac{\alpha^2}{\omega^2}(k+\beta) + \tfrac{\Lambda}{3}\big)\Big)\Big), \\ b_4 &= -\big(\tfrac{\alpha^2}{\omega^2}k + \tfrac{\Lambda}{3}\big) \end{aligned} \tag{2.4}$$

and

$$k = \frac{1 + 2\frac{\alpha}{\omega}\ell m - 3\ell^2\big(\frac{\alpha^2}{\omega^2}\beta + \frac{\Lambda}{3}\big)}{1 + 3\frac{\alpha^2}{\omega^2}\ell^2(a^2 - \ell^2)}(a^2 - \ell^2), \qquad \omega = \sqrt{a^2 + \ell^2}. \tag{2.5}$$

Basically, the coordinates t and r may range over all of $\mathbb{R}$ while ϑ and φ are standard coordinates on the two-sphere. Note, however, that for some values of the black-hole parameters r and ϑ have to be restricted, see Sect. 2.3. The Plebański–Demiański space-time depends on seven parameters (m, a, β, ℓ, α, Λ and C) which are to be interpreted in the following way: m is the mass of the black hole and a is its spin. β is a parameter that comprises electric and magnetic charge, $\beta = q_e^2 + q_m^2$ at least, if it is non-negative; for negative β, the metric cannot be interpreted as a solution to the Einstein–Maxwell equations because the electric or magnetic charge has to be imaginary then. Nonetheless, the case $\beta < 0$ is of interest because metrics of this form occur in some braneworld scenarios (Aliev and Gümrükçüoğlu 2005). The NUT parameter ℓ is to be interpreted as a gravitomagnetic charge (Griffiths and Podolský 2009, p. 219). The parameter α gives the acceleration of the black hole (Griffiths and Podolský 2009, p. 258) while Λ is the cosmological constant. The quantity C, which was introduced by Manko and Ruiz (2005), is relevant only if

$\ell \neq 0$. In this case, there is an *axial* singularity on the z axis and by choosing C appropriately this singularity can be distributed symmetrically or asymmetrically on the positive and the negative z axis. All the parameters, m, a, ℓ, β, Λ, α and C, may take arbitrary real values in principle, albeit not all possibilities are physically relevant.

If only the mass and the acceleration parameter are different from zero, we have the so-called C-metric[2] which describes a space-time with boost-rotation symmetry. This solution to the vacuum Einstein field equation was found by Levi-Civita (1919) and Weyl (1917, 1919). The name C-metric refers to the classification in the review of Ehlers and Kundt (1962). The rotating version of the C-metric was considered by Hong and Teo (2005) while a detailed discussion of accelerated space-times in general can be found in the book by Griffiths and Podolský (2009).

Commonly the C-metric is given in the form introduced by Hong and Teo (2003)

$$g^C_{\mu\nu}\,\mathrm{d}x^\mu\,\mathrm{d}x^\nu = \frac{1}{\alpha^2(x+y)^2}\left(-F(y)\,\mathrm{d}\tau^2 + \frac{\mathrm{d}y^2}{F(y)} + \frac{\mathrm{d}x^2}{G(x)} + G(x)\,\mathrm{d}\varphi^2\right) \tag{2.6}$$

with cubic functions $F(y) = -(1-y^2)(1-2\alpha m y)$ and $G(x) = (1-x^2)(1+2\alpha m x)$. The metric depends on two parameters, the mass m and the acceleration parameter α. The domain covered by the coordinates (τ, x, y, φ) actually contains *two* black holes accelerating away from each other with a conical singularity (a “strut”) on the axis of rotational symmetry (Griffiths and Podolský 2009; Kinnersley and Walker 1970; Bonnor 1983; Bonnor and Davidson 1992). For our purposes, Boyer–Lindquist coordinates are more suitable, see Eq. (2.1), which cover only one of the two black holes.

The Plebański–Demiański class (2.1) covers many well-known space-times like the Schwarzschild, Kerr or Taub–NUT space-time; their charged versions ($\beta > 0$) and versions with non-vanishing cosmological constant Λ or acceleration α are also included. The non-accelerated space-times ($\alpha = 0$) are comprised in the Plebański or Kerr–Newman–NUT–(anti-)de Sitter class of metrics. Details about the covered space-times and the particular parameters of the space-times can be found in Table 2.1; a similar one is also presented in the book by Stephani et al. (2003, p. 325).

In some of these cases, the two polynomials Δ_ϑ and Δ_r, see (2.2), reduce to much simpler forms. For $\alpha = 0$, we find $k = (1-\ell^2\Lambda)(a^2-\ell^2)$, $\omega = \sqrt{a^2+\ell^2}$ and hence

$$\begin{aligned}\Delta_\vartheta &= 1 + \Lambda\left(\tfrac{4}{3}a\ell\cos\vartheta + \tfrac{1}{3}a^2\cos^2\vartheta\right),\\ \Delta_r &= \Delta - \Lambda\left((a^2-\ell^2)\ell^2 + (\tfrac{1}{3}a^2+2\ell^2)r^2 + \tfrac{1}{3}r^4\right),\end{aligned} \tag{2.7}$$

while $\ell = 0$ yields $k = a^2$, $\omega = \sqrt{a^2+\ell^2} = a$ and

[2] Note that this parameter C has nothing to do with the name “C-metric” for space-times of accelerated black hole(s).

Table 2.1 Metrics covered in the Plebański–Demiański class

a	β	ℓ	α	Λ	Space-time
×	×	×	×	×	Plebański–Demiański
					Schwarzschild
	×				Reissner–Nordström
×					Kerr
×	×				Kerr–Newman
		×			Taub–NUT
×		×			Kerr–NUT
×	×	×			Kerr–Newman–NUT
				×	Kottler or Schwarzschild–(anti-)de Sitter
×	×	×		×	Plebański or Kerr–Newman–NUT–(anti-)de Sitter
			×		C-metric or accelerated Schwarzschild
×			×		Rotating C-metric or accelerated Kerr

The × marks the particular black hole parameters of the space-time additional to the mass m

$$\begin{aligned}\Delta_\vartheta &= 1 - 2\alpha m \cos\vartheta + \left(\alpha^2(a^2+\beta) + \tfrac{\Lambda}{3}a^2\right)\cos^2\vartheta,\\ \Delta_r &= \Delta(1-\alpha^2 r^2) - \tfrac{\Lambda}{3}(a^2+r^2)r^2,\end{aligned} \tag{2.8}$$

where $\Delta = r^2 - 2mr + a^2 - \ell^2 + \beta$.

2.1 Symmetries

In the Plebański–Demiański class, all metric coefficients $g_{\mu\nu}$ noted in Eq. (2.1) are independent of t and φ which is why all space-times of this class stay invariant under translations of t and φ. Thus, the corresponding coordinate vector fields

$$\partial_t = \tfrac{\partial}{\partial t}, \qquad \partial_\varphi = \tfrac{\partial}{\partial\varphi} \tag{2.9}$$

are Killing vector fields that describe the symmetries of the space-time. Since their scalar products reproduce the t and φ metric coefficients

$$g_{tt} = g(\partial_t, \partial_t), \qquad g_{t\varphi} = g(\partial_t, \partial_\varphi), \qquad g_{\varphi\varphi} = g(\partial_\varphi, \partial_\varphi) \tag{2.10}$$

these coefficients have a geometric meaning, see Sect. 2.4.

2.2 Isometries

We have learned from the symmetries that it is not important at which time or at which angle φ we are looking at the black hole. The metric is the same and consequently also the described geometry. Of course, the translations of t and φ are isometries but there are more. In general, space-times which differ in one of the black hole parameters describe different geometric situations since the metric is changed. But for the Plebański–Demiański class there are globally isometric cases with opposite signs for some black hole parameters. Two are given by the coordinate transformations, see Appendix B

$$\begin{array}{ccc} (M_{[m,\ a,\beta,\ell,\ C,\alpha,\lambda]}, g) & & (M_{[m,a,\beta,\ \ell,\ C,\ \alpha,\lambda]}, g) \\ \Big\downarrow \binom{\vartheta}{\varphi} \mapsto \binom{\pi-\vartheta}{-\varphi} & & \Big\downarrow \binom{\vartheta}{\varphi} \mapsto \binom{\pi-\vartheta}{\varphi} \\ (M'_{[m,-a,\beta,\ell,-C,\alpha,\lambda]}, g') & & (M'_{[m,a,\beta,-\ell,-C,-\alpha,\lambda]}, g') \end{array} \tag{2.11}$$

These isometries tell us the following: Two black holes ($C = 0$) which differ in the rotation direction only describe the same geometry but space-times have to be mirrored at the equatorial plane ($\vartheta \mapsto \pi - \vartheta$) and at the plane defined by the rotation axis and the $\varphi = 0$ direction ($\varphi \mapsto -\varphi$). For black holes which differ in the sign of ℓ and α, one gets space-times mirrored at the equatorial plane.

2.3 Singularities

The metric (2.1) becomes singular at the roots of Ω, Σ, Δ_r, Δ_ϑ and $\sin\vartheta$. Some of them are mere coordinate singularities while others are true (curvature) singularities. As this issue is of some relevance for our purpose, we briefly discuss the different types of singularities in the following paragraphs.

Conformal factor. Ω becomes zero if

$$r = \frac{\sqrt{a^2 + \ell^2}}{\alpha(\ell + a\cos\vartheta)}. \tag{2.12}$$

As the metric blows up if $\Omega \to 0$, Eq. (2.12) determines the boundary of the space-time. Because Ω enters as square into the metric (2.1), we restrict the space-time without loss of generality to the region where Ω is positive, see Fig. 2.4 on page 29. The allowed region is a half-space bounded by a plane ($\ell = 0$), a half-space bounded by one sheet of a two-sheeted hyperboloid ($\ell^2 < a^2$), a domain bounded by a paraboloid ($\ell^2 = a^2$), or a domain bounded by an ellipsoid ($\ell^2 > a^2$); see Fig. 2.1 for appropriate illustrations of these regions. For $\alpha = 0$ there is no restriction because $\Omega \equiv 1$. Note that Eq. (2.12) gives positive as well as negative r values depending on the signs of α, ℓ, a.

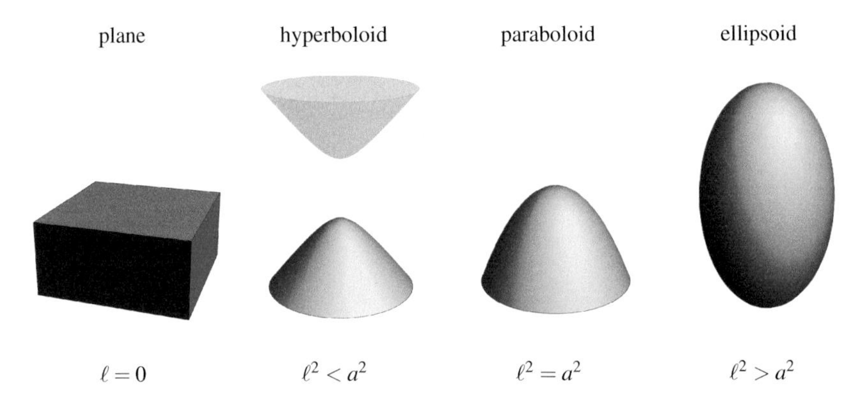

Fig. 2.1 Hyperboloids

Ring singularity. The equation $\Sigma = 0$ is equivalent to

$$r = 0 \quad \text{and} \quad \cos\vartheta = -\ell/a. \tag{2.13}$$

If $\ell^2 < a^2$, this condition is satisfied on a ring. The singularity on this ring turns out to be a true (curvature) singularity if $m \neq 0$. It is usually referred to as the *ring singularity*. Note that, apart from this singularity, the sphere $r = 0$ is regular. Hence, it is possible to travel through either of the two hemispheres ("throats") that are bounded by the ring singularity—from the region $r > 0$ to the region $r < 0$ or vice versa.

If $\ell^2 > a^2$, there is no ring singularity. Σ is different from zero everywhere and the entire sphere $r = 0$ is regular.

In the limiting case where $\ell^2 = a^2$, the ring singularity degenerates into a point on the axis. It becomes a point singularity for $\ell = a = 0$ that disconnects the space-time into the regions $r > 0$ and $r < 0$. The ring singularity is unaffected by α.

Axial singularity. The metric is singular on the z axis, i.e. for $\sin\vartheta = 0$, and this is always the case when using spherical polar coordinates. If $\alpha \neq 0$ or $\ell \neq 0$, this is not just a coordinate singularity but rather a true (conical) singularity on (at least a part of) the rotational axis. In the NUT case, the singularity depends on the Manko–Ruiz parameter C.

To demonstrate this, we observe that in the limit $\cos\vartheta \to \pm 1$ we have $\Sigma \to r^2 + (\ell \pm a)^2$ and $\chi \to -2\ell(\pm 1 + C)$. As a consequence, the metric coefficient

$$g^{tt} = \Omega^2 \left(\frac{\chi^2}{\Sigma \Delta_\vartheta \sin^2\vartheta} - \frac{(\Sigma + a\chi)^2}{\Sigma \Delta_r} \right) \tag{2.14}$$

diverges unless $C = \mp 1$. This divergent behavior indicates that either the coordinate function t or the metric g becomes pathological. It was shown by Misner (1963) that this singularity can be removed if one makes the time coordinate t periodic.

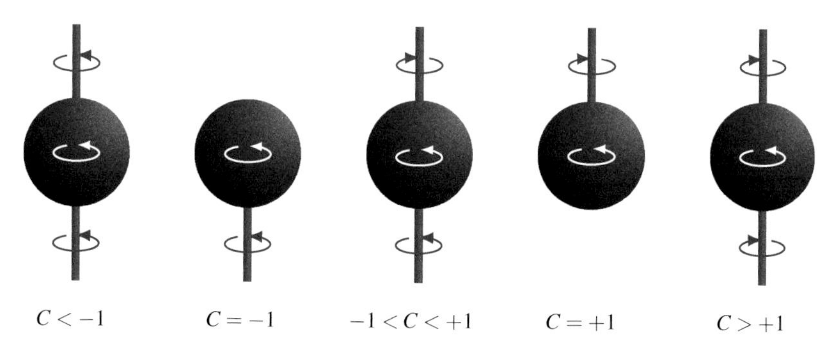

Fig. 2.2 Singularities on the z axis in Kerr–NUT space-times marked as (*red*) rotating rods

(Misner restricted himself to the Taub–NUT metric, $a = \beta = \Lambda = 0$, with $C = 1$ but his reasoning applies equally well to the general case.) We do *not* follow this suggestion because it leads to a space-time with closed timelike curves through *every* event. Instead, we adopt Bonnor's interpretation (Bonnor 1969, p. 145) of the axial singularity who viewed it as a "massless source of angular momentum", see also Stephani et al. (2003, p. 310). As pointed out by Manko and Ruiz (2005), this source term is splitted into two semi-infinite rotating rods with negative masses and infinite angular momenta where the rotation direction of the rods depends on C, see Fig. 2.2. The Manko–Ruiz parameter C is balancing the singularity. For $C = 1$, the singularity is on the half-axis $\vartheta = 0$, for $C = -1$ it is on the half-axis $\vartheta = \pi$ and for any other value of C it is on both half-axes. Thus, by choosing the Manko–Ruiz parameter C appropriately, one can decide on which part of the axis the singularity is situated. Note that each half-axis extends from $r = -\infty$ to $r = \infty$.

Metrics (2.1) with different values of C are *locally* isometric near all points off the axis. This follows from the fact that a coordinate transformation $t' = t - 2\ell\widetilde{C}\varphi$ yields, again, a metric (2.1) with $C' = C + \widetilde{C}$. For $\widetilde{C} = -C$ such a coordinate transformation eliminates the parameter C from the metric, see Kagramanova et al. (2010). Note, however, that this transformation does not work globally because φ is periodic and t is not, and it does not work near the axis because φ is pathological there. But there are globally isometric space-times as pointed out in Sect. 2.2.

Horizons. Straumann (2013, p. 471ff) explained in detail that horizons for the Kerr–Newman family are the null hypersurfaces

$$H = \{g(\xi, \xi) = 0\} \qquad \text{with } \xi = \partial_t - \frac{g_{t\varphi}}{g_{\varphi\varphi}}\partial_\varphi. \tag{2.15}$$

His argumentation applies equally well to the general Plebański–Demiański class. Therefore, the horizons of a Plebański–Demiański black hole are at

$$g(\xi, \xi) = 0 \qquad \Longleftrightarrow \qquad 0 = g_{\varphi t}^2 - g_{tt}g_{\varphi\varphi} = \frac{\Delta_r \Delta_\vartheta \sin^2\vartheta}{\Omega^4} \tag{2.16}$$

and can be found as real roots of Δ_r or Δ_ϑ which are coordinate singularities. In the following, we successively discuss both cases.

(i) A Plebański–Demiański space-time exhibits up to 4 horizons $r_1 > r_2 > \ldots$ on spheres $r =$ constant since Δ_r is in general a polynomial of degree 4. If $\alpha = 0$ and $\Lambda = 0$, then Δ_r reduces to a second-degree polynomial

$$\Delta_r = r^2 - 2mr + a^2 - \ell^2 + \beta \tag{2.17}$$

and horizons can be found at

$$r_\pm = m \pm \sqrt{m^2 - a^2 + \ell^2 - \beta} \tag{2.18}$$

as long as $a^2 \leq a_{\max}^2 := m^2 + \ell^2 - \beta$. Then $r_+ (= r_1)$ is the outer (event) horizon of the black hole and $r_- (= r_2)$ is the inner horizon. For $a^2 > a_{\max}^2$ we would find, instead of a black hole, a naked singularity or a regular space-time. But we will not consider this possibility in the following because we are interested in the black hole case only. Then, the spin a is bounded by $a_{\max}$ and a maximally rotating black hole ($a^2 = a_{\max}^2$) is called *extremal black hole*. Since ∂_r is space like outside of the event horizon ($\Delta_r > 0$), communication is possible here. Therefore, this region is called *domain of outer communication* and we will place our observers for observing the shadow of the black hole within this region.

With cosmological constant (but $\alpha = 0$) we obtain for Δ_r, see Eq. (2.7)

$$\Delta_r = (r^2 - 2mr + a^2 - \ell^2 + \beta) - \Lambda\big((a^2 - \ell^2)\ell^2 + (\tfrac{1}{3}a^2 + 2\ell^2)r^2 + \tfrac{1}{3}r^4\big) \tag{2.19}$$

which has a strictly positive second derivative Δ_r'' if $\Lambda < 0$, as for $\Lambda = 0$. Hence, the number of zeros of Δ_r is either 2 or 0 and as above we have a black hole or a naked singularity or regular space-time. Again, the domain of outer communication around the black hole is the region between $r = \infty$ and the first horizon at r_1. If $\Lambda > 0$, the vector field ∂_r is timelike for big values of r. Therefore, the first horizon, if existing, is a cosmological horizon. We have a black hole if there are four horizons altogether. Then, the domain of outer communication is the region between the first and the second horizon. But in both cases the horizons could in general not be specified in a simple form because of the higher degree of Δ_r.

This applies also to an accelerated scenario with nonvanishing NUT charge ℓ or cosmological constant Λ. But if both are zero, the horizons can easily be determined. According to Eq. (2.8) Δ_r is factorized then

$$\Delta_r = (r^2 - 2mr + a^2 + \beta)(1 - \alpha^2 r^2); \tag{2.20}$$

therefore, we find the usual (Kerr–Newman) horizons at $r = r_\pm$ given by Eq. (2.18) with $\ell = 0$ and additional cosmological horizons at $r = \pm\frac{1}{\alpha}$. Of course, we must

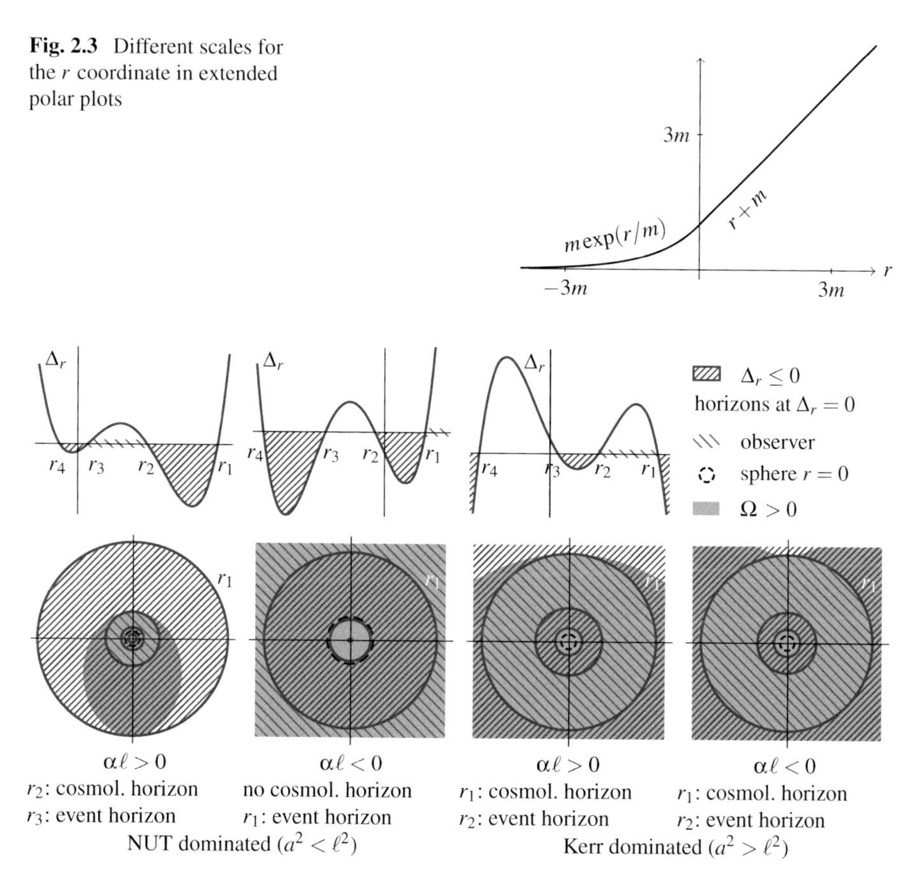

Fig. 2.3 Different scales for the r coordinate in extended polar plots

Fig. 2.4 Schematic illustrations of the graph of Δ_r (*upper* row) and extended polar plots of the region $\Omega > 0$ (*lower* row). Depending on the sign of its leading coefficient, Δ_r goes to $\pm\infty$ for big radii r; the sign changes if $a^2 \approx \ell^2$ (equality for $\Lambda = 0$) and with the change the space-time is no longer NUT but Kerr dominated. The space-time is restricted to that region where $\Omega > 0$ ( ). Geometrically, the boundary of this region is an ellipsoid (*left*) or one sheet of a two-sheeted hyperboloid (*right*). In the NUT dominated case, the root r_1 of Δ_r is not contained in the ellipsoid $\Omega > 0$ for $\alpha\ell > 0$; thus, the event horizon is at r_3 instead. Interestingly, for $\alpha\ell < 0$ there is no cosmological horizon. The *red* hatched region ( ) marks the outer domain of communication ($\Delta_r > 0$) where observers are placed

have $|\alpha| < \frac{1}{r_+}$. For analyzing the general case, we have to take into account that only the regions where $\Omega > 0$, see Eq. (2.12), are allowed. Here, the vector field ∂_r could be timelike or space like for big values of r and this depends, of course, on the sign of the leading coefficient b_4 of Δ_r. Figure 2.4 shows in the first row schematic illustrations of the graph of Δ_r and in the second row *extended* polar plots of the region $\Omega > 0$.

Following a suggestion by O'Neill (1995), we show the entire range of the space-time, with the Boyer–Lindquist coordinate r increasing outward from the origin which corresponds to $r = -\infty$. But in order to not highlight the outer parts by a strong deformation, we use two different scales for the radial coordinate (see Fig. 2.3): In the inner region $r < 0$ (inside the sphere $r = 0$ marked by a dashed circle), the radial coordinate is plotted as $m \exp(r/m)$; this is continuously extended with $r + m$ in the outer region $r > 0$ (outside the sphere $r = 0$).

The sign of b_4 does not only define the causal character of ∂_r but also the character of the whole space-time since the sign changes at $a^2 \approx \ell^2$ (equality for $\Lambda = 0$), see Eq. (2.4). Thus, the space-time is NUT dominated with space like ∂_r ($b_4 > 0$) for big r if $a^2 < \ell^2$ and Kerr dominated with timelike ∂_r ($b_4 < 0$) for big r if $a^2 > \ell^2$.

In the NUT case (left columns in Fig. 2.4) where according to Eq. (2.13) we have no ring singularity, interesting things happen. For $\alpha\ell > 0$, the first root r_1 is not in the allowed region $\Omega > 0$. Hence, we have a cosmological horizon at r_2, the event horizon of the black hole at r_3 and in between the domain of outer communication. Different signs of α and ℓ, however, result in 3 negative roots of Δ_r; thus, the only positive root r_1 is the event horizon and the adjacent domain of outer communication is not bounded by a cosmological horizon. However, one can easily read of Eq. (2.12) that $\Omega < 0$ in the equatorial plane ($\vartheta = \frac{\pi}{2}$) is only possible for negative r values. The timelike case (right columns in Fig. 2.4) is similar to the non-accelerated space-times discussed before. Since all real roots of Δ_r are in the allowed region with $\Omega > 0$, the first root r_1 represents a cosmological horizon and the subsequent root r_2 is the black-hole horizon. Here, the domain of outer communication is the region between r_1 and r_2 where $\Delta_r > 0$.

(ii) As mentioned on page 28, the roots of Δ_ϑ are coordinate singularities, too; these indicate further horizons where the vector field ∂_ϑ changes the causal character from space like to timelike, just as the vector field ∂_r does at the roots of Δ_r. However, since these horizons lie on cones $\vartheta =$ constant instead of spheres $r =$ constant, such a situation would be hardly of any physical relevance. Therefore, we exclude it by limiting the parameters of the black hole appropriately: As $\Delta_\vartheta = 0$ implies

$$\cos\vartheta_\pm = \frac{-a_3 \pm \sqrt{a_3^2 + 4a_4}}{2a_4}, \tag{2.21}$$

$\Delta_\vartheta \neq 0$ is guaranteed for all real ϑ if the radicand in Eq. (2.21) is negative or if the absolute value of the entire right hand side of (2.21) is greater than 1. In all subsequently considered cases one of these sufficient conditions is fulfilled, see the corresponding tables in Appendix A.

For some special cases these conditions define simple constraints on the parameters. If $\alpha = 0$, Eq. (2.21) ends in $a \cos\vartheta_\pm = -2\ell \pm \sqrt{4\ell^2 - 3/\Lambda}$ and

$$a_3^2 + 4a_4 < 0 \qquad \Longleftrightarrow \qquad 4\ell^2\Lambda < 3. \tag{2.22}$$

For $\ell = 0$, $\Lambda = 0$ we find

$$\alpha \cos \vartheta_{\pm} = \frac{m \pm \sqrt{m^2 - \beta - a^2}}{a^2 + \beta} \geq \frac{1}{2m} \tag{2.23}$$

since $a^2 \leq a_{\max}^2 = m^2 - \beta$ and $\beta < m^2$. Thus, $\Delta_\vartheta \neq 0$ is assured if $|\alpha| < \frac{1}{2m}$.

2.4 Ergoregion and Causality Violation

There are some other interesting regions around a black hole characterized by the change of the causal character of the Killing vector fields ∂_t and ∂_φ.

In the region where ∂_t becomes space like, i.e. $g_{tt} > 0$, no observer can move on a t-line. Thus, any observer in this region has to rotate (in φ direction). This region with $g_{tt} > 0$ is known as the *ergosphere* or the *ergoregion*,[3]

$$g_{tt} = -\frac{1}{\Omega^2 \Sigma}\left(\Delta_r - a^2 \Delta_\vartheta \sin^2 \vartheta\right); \tag{2.24}$$

its boundary, $g_{tt} = 0$, is called *static limit*. An ergoregion only exists if $a \neq 0$. Note that at the horizons, i.e., at the roots of Δ_r, the metric coefficient g_{tt} is positive, see Eq. (2.24). Hence, the horizons are always contained within the ergoregion. For $\alpha \neq 0$ or $\Lambda \neq 0$ there are cosmological horizons in addition to the black hole horizons; then the ergoregion consists of several connected components. At the poles ($\vartheta = 0, \pi$), the boundary of the ergoregion and the horizon share a common tangential plane. Since $\sin(\frac{\pi}{2} + \vartheta) = \sin(\frac{\pi}{2} - \vartheta)$, the ergoregion is symmetric with respect to the equatorial plane if $\alpha = 0$ and $\Lambda = 0$, cf. Eqs. (2.24) and (2.7). Actually, the ergoregion stays almost symmetric for small values of Λ, $\Lambda \leq 10^{-2}\text{m}^{-2}$, because then $\Delta_\vartheta \approx 1$. This behavior is lost with an acceleration $\alpha \neq 0$.

If $a \neq 0$ or $\ell \neq 0$, there are regions where the Killing field ∂_φ becomes timelike, $g_{\varphi\varphi} < 0$. This indicates causality violation because the φ-lines are closed timelike curves. For $\ell = 0$, the region where $g_{\varphi\varphi} = 0$ is completely contained in the domain where $r < 0$ and, thus, hidden behind the horizon for an observer in the domain of outer communication. In the case $\ell \neq 0$, however, there is a causality violating region in the domain of outer communication around the axial singularity.

By [1–3] I refer to my papers Grenzebach et al. (2014), Grenzebach (2015) and Grenzebach et al. (2015), respectively. Sentences marked with [i] can be found in total or only slightly modified in the ith paper

[3] Some authors call only the region between the horizon and the static limit *ergoregion*. This is that part of the region $g_{tt} > 0$ which an outside observer would be able to see.

References

Aliev AN, Gümrükçüoğlu AE (2005) Charged rotating black holes on a 3-brane. Phys Rev D 71(10):104,027(14). doi:10.1103/PhysRevD.71.104027

Bonnor WB (1969) A new interpretation of the NUT metric in general relativity. Math Proc Cambridge Philos Soc 66(1):145–151. doi:10.1017/S0305004100044807

Bonnor WB (1983) The sources of the vacuum C-Metric. Gen Relativ Gravit 15(6):535–551. doi:10.1007/BF00759569

Bonnor WB, Davidson W (1992) Interpreting the Levi-Civita vacuum metric. Class Quantum Gravity 9(9):2065–2068. doi:10.1088/0264-9381/9/9/012

Carter B (1968) Hamilton-Jacobi and Schrödinger separable solutions of Einstein's equations. Commun Math Phys 10(4):280–310. http://projecteuclid.org/euclid.cmp/1103841118

Debever R (1971) On type D expanding solutions of Einstein-Maxwell equations. Bulletin de la Société Mathématique de Belgique 23:360–376

Ehlers J, Kundt W (1962) Exact solutions of the gravitational field equations. In: Witten, chap 2, pp 49–101

Grenzebach A, Perlick V, Lämmerzahl C (2014) Photon regions and shadows of Kerr–Newman–NUT Black Holes with a cosmological constant. Phys Rev D 89:124,004(12). doi:10.1103/PhysRevD.89.124004. arXiv:1403.5234

Grenzebach A (2015) Aberrational effects for shadows of black holes. In: Puetzfeld et al Proceedings of the 524th WE-Heraeus-Seminar "Equations of Motion in Relativistic Gravity", held in Bad Honnef, Germany, 17–23 Feb 2013, pp 823–832. doi:10.1007/978-3-319-18335-0_25, arXiv:1502.02861

Grenzebach A, Perlick V, Lämmerzahl C (2015) Photon regions and shadows of accelerated black holes. Int J Mod Phys D 24(9):1542,024(22). doi:10.1142/S0218271815420249 ("Special Issue Papers" of the "7th Black Holes Workshop", Aveiro, Portugal, arXiv:1503.03036)

Griffiths JB, Podolský J (2009) Exact space-times in Einstein's general relativity. Cambridge Monographs on Mathematical Physics, Cambridge University Press, Cambridge. doi:10.1017/CBO9780511635397

Hong K, Teo E (2003) A new form of the C-metric. Class Quantum Gravity 20(14):3269–3277. doi:10.1088/0264-9381/20/14/321

Hong K, Teo E (2005) A new form of the rotating C-metric. Class Quantum Gravity 22(1):109–117. doi:10.1088/0264-9381/22/1/007

Kagramanova V, Kunz J, Hackmann E, Lämmerzahl C (2010) Analytic treatment of complete and incomplete geodesics in Taub–NUT space-times. Phys Rev D 81(12):124,044(17). doi:10.1103/PhysRevD.81.124044

Kinnersley W, Walker M (1970) Uniformly Accelerating Charged Mass in General Relativity. Phys Rev D 2(8):1359–1370. doi:10.1103/PhysRevD.2.1359

Levi-Civita T (1919) ds^2 einsteiniani in campi newtoniani. VIII. Soluzioni binarie de Weyl. Rendiconti della Reale Accademia dei Lincei 28(1):3–13

Manko VS, Ruiz E (2005) Physical interpretation of the NUT family of solutions. Class Quantum Gravity 22(17):3555–3560. doi:10.1088/0264-9381/22/17/014

Miller JG (1973) Global analysis of the KerrTaubNUT metric. J Math Phys 14(4):486–494. doi:10.1063/1.1666343

Misner CW (1963) The flatter regions of Newman, Unti, and Tamburino's generalized Schwarzschild space. J Math Phys 4(7):924–937. doi:10.1063/1.1704019

O'Neill B (1995) The geometry of Kerr black holes. A K Peters, Wellesley

Plebański JF, Demiański M (1976) Rotating, charged and uniformly accelerating mass in general relativity. Ann Phys 98(1):98–127. doi:10.1016/0003-4916(76)90240-2

Stephani H, Kramer D, MacCallum M, Hoenselaers C, Herlt E (2003) Exact solutions of Einstein's field equations, 2nd edn. Cambridge Monographs on Mathematical Physics, Cambridge University Press, New York. doi:10.1017/CBO9780511535185

Straumann N (2013) General Relativity, 2nd edn. Graduate texts in physics, Springer, Dordrecht. doi:10.1007/978-94-007-5410-2

Weyl H (1919) Raum, Zeit, Materie, 3rd edn. Springer, Berlin. http://www.archive.org/details/raumzeitmateriev00weyl

Weyl H (1917) Zur Gravitationstheorie. Annalen der Physik 359(18):117–145. doi:10.1002/andp.19173591804

Chapter 3
Photon Regions Around Black Holes

Abstract The existence of a photon region, a region that contains spherical light-like geodesics, is crucial for determining the shadow of a black hole. Here, their characterizing inequality is derived. The photon regions are visualized together with ergoregions and regions with causality violation for various values of the parameters.

Keywords Photon region black hole · Equations of motion · Spherical light-rays · Plots photon region · Spin · Charge · NUT charge · Cosmological constant · Accelerated space-time

In the Plebański class of space-times, i.e., for $\alpha = 0$, the geodesic equation is completely integrable.[1] In addition to the obvious constants of motion, there is a fourth constant of motion, known as the Carter constant, which is associated with a second-rank Killing tensor. If $\alpha \neq 0$, instead of this Killing tensor we only have a conformal Killing tensor. This is sufficient to assure complete integrability for *lightlike* geodesics. The four constants of motion in a Plebański–Demiański space-time are the Lagrangian

$$\mathscr{L} = \tfrac{1}{2} g_{\mu\nu}\dot{x}^{\mu}\dot{x}^{\nu}, \qquad \mathscr{L} = 0 \quad \text{for light,} \tag{3.1}$$

(the dot denotes a derivative with respect to an affine parameter τ), the energy E and the z-component L_z of the angular momentum

$$E = -\frac{\partial \mathscr{L}}{\partial \dot{t}} = -g_{\varphi t}\dot{\varphi} - g_{tt}\dot{t}, \qquad L_z = \frac{\partial \mathscr{L}}{\partial \dot{\varphi}} = g_{\varphi\varphi}\dot{\varphi} + g_{\varphi t}\dot{t}, \tag{3.2}$$

which are associated with the Killing vectors ∂_t and ∂_φ from Sect. 2.1 and the Carter constant K, see Carter (1968). The last-mentioned Carter constant may be regarded as the separation constant for the r and the ϑ motion of lightlike geodesics.[2] The

[1] Large parts of this section are based on [1, 3] while parts of the Sects. 3.1, 3.2 and 3.3 can be found in [1] or [3].

[2] There are other definitions of the Carter constant which differ in additive constants.

A. Grenzebach, *The Shadow of Black Holes*,
SpringerBriefs in Physics, DOI 10.1007/978-3-319-30066-5_3

four constants of motion allow us to write the lightlike geodesic equation, $\mathscr{L} = 0$, in separated first-order form[3]

$$\frac{\Sigma}{\Omega^2}\dot{t} = \frac{\chi(L_z - E\chi)}{\Delta_\vartheta \sin^2\vartheta} + \frac{(\Sigma + a\chi)\big((\Sigma + a\chi)E - aL_z\big)}{\Delta_r}, \tag{3.3a}$$

$$\frac{\Sigma}{\Omega^2}\dot{\varphi} = \frac{L_z - E\chi}{\Delta_\vartheta \sin^2\vartheta} + \frac{a\big((\Sigma + a\chi)E - aL_z\big)}{\Delta_r}, \tag{3.3b}$$

$$\left(\frac{\Sigma}{\Omega^2}\right)^2 \dot{\vartheta}^2 = \Delta_\vartheta K - \frac{(\chi E - L_z)^2}{\sin^2\vartheta} =: \Theta(\vartheta), \tag{3.3c}$$

$$\left(\frac{\Sigma}{\Omega^2}\right)^2 \dot{r}^2 = \big((\Sigma + a\chi)E - aL_z\big)^2 - \Delta_r K =: R(r). \tag{3.3d}$$

The equations of motion can be solved explicitly in terms of hyperelliptic functions (Hackmann et al. 2009; Hackmann 2010; Kagramanova et al. 2010). Here, we are interested in spherical lightlike geodesics, i.e., lightlike geodesics that stay on a sphere $r =$ constant. The region filled by these geodesics is called the *photon region* $\mathscr{K}$. Mathematically, spherical orbits are characterized by $\dot{r} = 0$ and $\ddot{r} = 0$ which requires by (3.3d) that $R(r) = 0$ and $R'(r) = 0$, where the prime stands for the derivative with respect to r. Thus

$$K_E = \frac{\big((\Sigma + a\chi) - aL_E\big)^2}{\Delta_r}, \qquad K_E = \frac{4r\big((\Sigma + a\chi) - aL_E\big)}{\Delta_r'}, \tag{3.4}$$

where K_E and L_E are abbreviations

$$K_E = \frac{K}{E^2}, \qquad L_E = \frac{L_z}{E}. \tag{3.5}$$

After solving (3.4) for the constants of motion

$$K_E = \frac{16r^2\Delta_r}{(\Delta_r')^2}, \qquad aL_E = \big(\Sigma + a\chi\big) - \frac{4r\Delta_r}{\Delta_r'}, \tag{3.6}$$

we can substitute these expressions into (3.3c). As the left-hand side of (3.3c) is non-negative, $0 \le \left(\frac{\Sigma}{\Omega^2}\right)^2\dot{\vartheta}^2$, we find an inequality that determines the photon region

$$\mathscr{K}: \big(4r\Delta_r - \Sigma\Delta_r'\big)^2 \le 16a^2r^2\Delta_r\Delta_\vartheta \sin^2\vartheta. \tag{3.7}$$

Of course, the equality sign defines the boundary of the photon region. Note that $\mathscr{K}$ is independent of the Manko–Ruiz parameter C. Furthermore, all calculations

[3]This is also possible for matter if $\alpha = 0$ since Σ has no term depending on both r and ϑ.

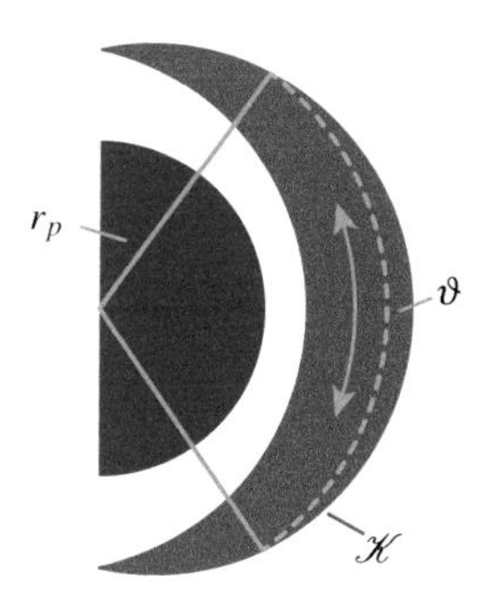

Fig. 3.1 Motion within the photon region $\mathscr{K}$

presented above do not differ between black holes or naked singularities. They are valid for both cases, yet.

Just as in the Kerr case (cf. Perlick 2004), through every point (r_p, ϑ_p) of $\mathscr{K}$ there is a lightlike geodesic that stays on the sphere $r = r_p$. Along each of these spherical lightlike geodesics, the ϑ motion is an oscillation bounded by the boundary of $\mathscr{K}$, see Fig. 3.1, while the φ motion given by (3.3b) might be quite complicated. For some spherical light rays it is not even monotonic. For pictures of individual spherical photon orbits around a Kerr black hole we refer to Teo (2003).

A non-rotating black hole ($a = 0$) is surrounded by a *photon sphere*, rather than by a photon region, since the inequality (3.7) defining $\mathscr{K}$ degenerates into an equality

$$4r\Delta_r = (r^2 + \ell^2)\Delta_r'. \tag{3.8}$$

The best known example is the photon sphere at $r = 3\,\mathrm{m}$ in the Schwarzschild space-time.

The stability of the spherical geodesic with respect to radial perturbations is determined by the sign of R''; a spherical geodesic at $r = r_p$ is unstable if $R''(r_p) > 0$, and stable if $R''(r_p) < 0$. The second derivative R'' can be calculated from (3.3d). With the help of (3.6) this results in

$$\frac{R''(r)}{8E^2}\Delta_r'^2 = 2r\Delta_r\Delta_r' + r^2\Delta_r'^2 - 2r^2\Delta_r\Delta_r''. \tag{3.9}$$

Because of the rotational symmetry, it is convenient to plot a meridional section through space-time for illustrating the regions around a black hole. The resulting pictures can be found in the following sections. They are extended (r, ϑ) polar diagrams where ϑ is measured from the positive z-axis. To view the full range of the space-times, the radial coordinate is plotted with a special scaling as introduced in Sect. 2.3 and depicted in Fig. 2.3 (exponentially for $r < 0$ and linearly for $r > 0$). Thus, $r = 0$ is a sphere whose throats are marked as dashed circle. Each figure contains the photon region $\mathscr{K}$, where unstable and stable spherical light rays are distinguished according to (3.9); the horizons $r_\pm$ of the black hole are given as

Fig. 3.2 Legend for all plots illustrating photon regions

- region with $\Delta_r \leq 0$; boundary ($\Delta_r = 0$): horizons
- unstable spherical light-rays in photon region $\mathscr{K}$
- stable spherical light-rays in photon region $\mathscr{K}$
- region with causality violation ($g_{\varphi\varphi} < 0$)
- ergoregion ($g_{tt} > 0$)
- throats at $r = 0$
- ring singularity

boundaries of the region where $\Delta_r \leq 0$. Furthermore, the ergoregion, the causality violating region and the ring singularity are shown. A legend for these quantities is given in Fig. 3.2.

3.1 Kerr–Newman–NUT Space-Times

The Figs. 3.4, 3.5 and 3.6 show meridional sections, i.e., extended (r, ϑ) diagrams in Kerr–Newman–NUT space-times ($\alpha = 0$, $\Lambda = 0$). We explore how the rotation a, the charge β or the gravitomagnetic NUT-charge ℓ of the black hole affects the nearby special regions. Each figure shows amongst others the photon region $\mathscr{K}$ for four different values of the spin a, keeping all the other parameters fixed, see Fig. 3.2 for a legend. Restricting to black-hole cases, we choose $a = \lambda a_{\max}$, where $\lambda \in \left\{\frac{1}{50}, \frac{2}{5}, \frac{4}{5}, 1\right\}$ and $a_{\max}$ denotes the spin of an extremal black hole which is determined by the other parameters $a_{\max} = \sqrt{m^2 + \ell^2 - \beta}$, cf. Eq. 2.18.

More precisely, each of the Figs. 3.4, 3.5 and 3.6 shows images of three space-times in the Kerr–Newman, Kerr–NUT, and Kerr–Newman–NUT class, respectively. The left column of every figure represents the Kerr space-time for the purpose of comparison since Kerr space-time is a subclass of each of these classes. Furthermore, the figures are splitted up into two subfigures where part (a) contains the hole range while part (b) shows a magnification of the inner part beyond the inner horizon r_-. Supplementary to the plots, all parameters of the black holes that are used in these three figures are listed in Table A.2 in the appendix together with the r values of the horizons.

In the **Kerr space-time**, as shown in the left columns of the figures, there is an exterior photon region at $r > r_+$ and an interior photon region at $r < r_-$. Both of them are symmetric with respect to the equatorial plane. Starting from the photon sphere at $r = 3\,\mathrm{m}$ for the non-rotating Schwarzschild case, the exterior photon region develops a crescent-shaped cross-section $\mathfrak{D}$ for $a \neq 0$ and grows with increasing spin a. The interior photon region consists of two connected components that are separated by the ring singularity. In the exterior photon region, all spherical light orbits are unstable while in the interior photon region there are stable and unstable ones. For a maximally rotating black hole, the exterior and interior photon regions touch the horizon.

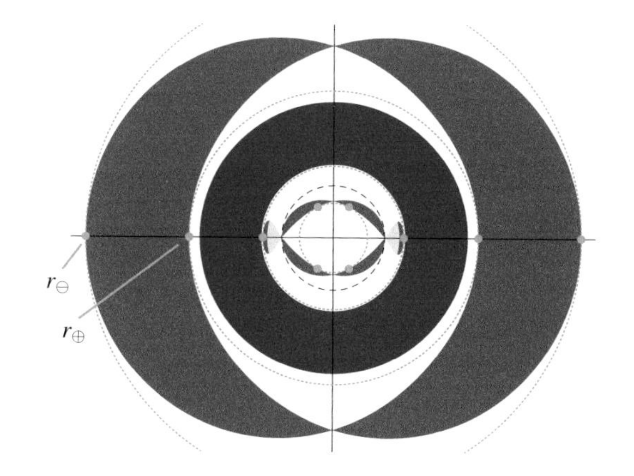

Fig. 3.3 Five circular photon orbits, marked with *green dots* on the boundary of the photon region $\mathscr{K}$ around a Kerr *black hole*. The corresponding spheres tangential to $\mathscr{K}$ are represented by the *dotted green circles*. The circular orbits at $r_\ominus$ and $r_\oplus$ are counter- and co-rotating, respectively, with the rotation of the *black hole*

As discussed earlier, a photon on an orbit in the photon region propagates in general on a sphere where the ϑ motion is an oscillation bounded by the boundary of the photon region. Only for a non-rotating black hole all these orbits are circular. But also a rotating black hole has circular photon orbits: A circular lightlike geodesic exists where the boundary of the photon region is tangential to a sphere $r =$ constant and is thus characterized by $\dot{\vartheta} = 0$ and $\ddot{\vartheta} = 0$. In Fig. 3.3, the circular orbits are marked with green dots in one exemplary plot. We easily recognize the three well-known circular lightlike geodesics in the equatorial plane, but also two not-so-well-known circular lightlike geodesics off the equatorial plane. The latter are situated in the region where $r < 0$. However, three of the orbits are hidden behind the inner horizon r_-; the other two are in the exterior photon region. Here, light rays on the outermost circular orbit with radius $r_\ominus$ are counter-rotating while those on the smaller orbit at $r_\oplus$ are co-rotating with the rotation of the black hole (cf. Hartle 2003).

The causality violating region is adjacent to the ring singularity in the Kerr space-time and lies to the side of negative r. Thus, this region is hidden behind the outer horizon similar to the three innermost circular photon orbits. For small a, the ergoregion does not intersect the exterior photon region, but for $a^2 > \frac{1}{2}m^2$ it does. This confirms that the circular photon orbit in the exterior photon region with smaller radius $r_\oplus$ a co-rotating orbit.

An added electric or magnetic charge parameter $\beta = q_e^2 + q_m^2$ of the **Kerr–Newman** space-time affects the photon regions little for small charges, see Fig. 3.4. For large charges, for instance $\beta = \frac{8}{9}m^2$, the exterior and interior photon regions do not touch the horizon any longer for a maximally rotating black hole. The reason for this is that the value of the maximal possible spin $a_{\max} = \sqrt{m^2 - \beta}$ of a maximally rotating black hole decreases with increasing charge β. Hence, some properties of highly charged and maximally rotating black holes are similar to those of slowly rotating but hardly charged black holes. Nevertheless, there is a qualitative effect of β: one of the two connected components of the interior photon region is now detached from the ring singularity.

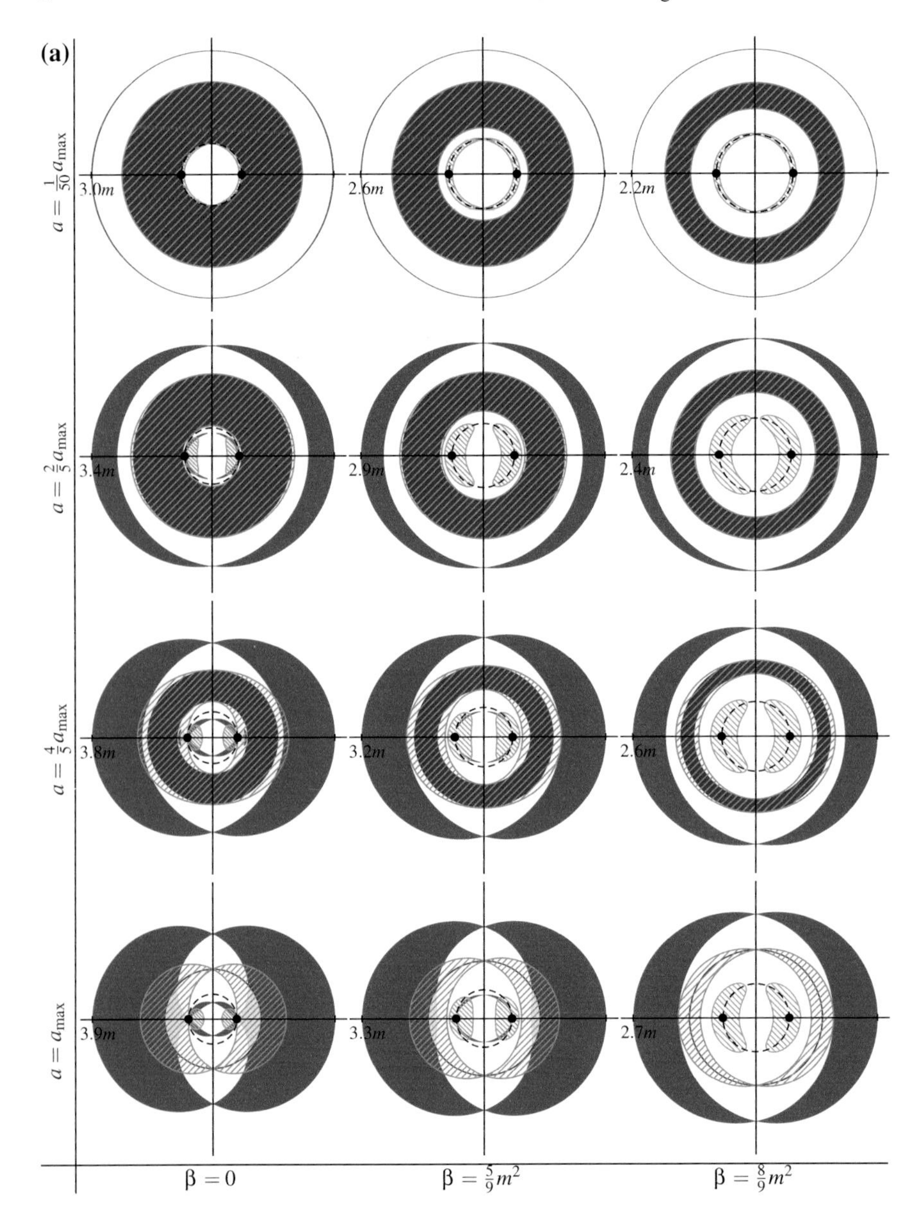

Fig. 3.4 Photon regions in Kerr–Newman space-times for spins $a = \lambda a_{\max}$ where $a_{\max} = \sqrt{m^2 - \beta}$. The *black hole* parameters are listed in Table A.2. **a** Photon regions in Kerr–Newman space-times. **b** Magnified inner part of plots in (**a**)

From Sect. 2.3 we know that a gravitomagnetic charge $\ell \neq 0$ results in a true singularity on the z axis. Hence, we have to expect different images for NUT space-times. At first, the additional gravitomagnetic charge ℓ of the **Kerr–NUT** space-time

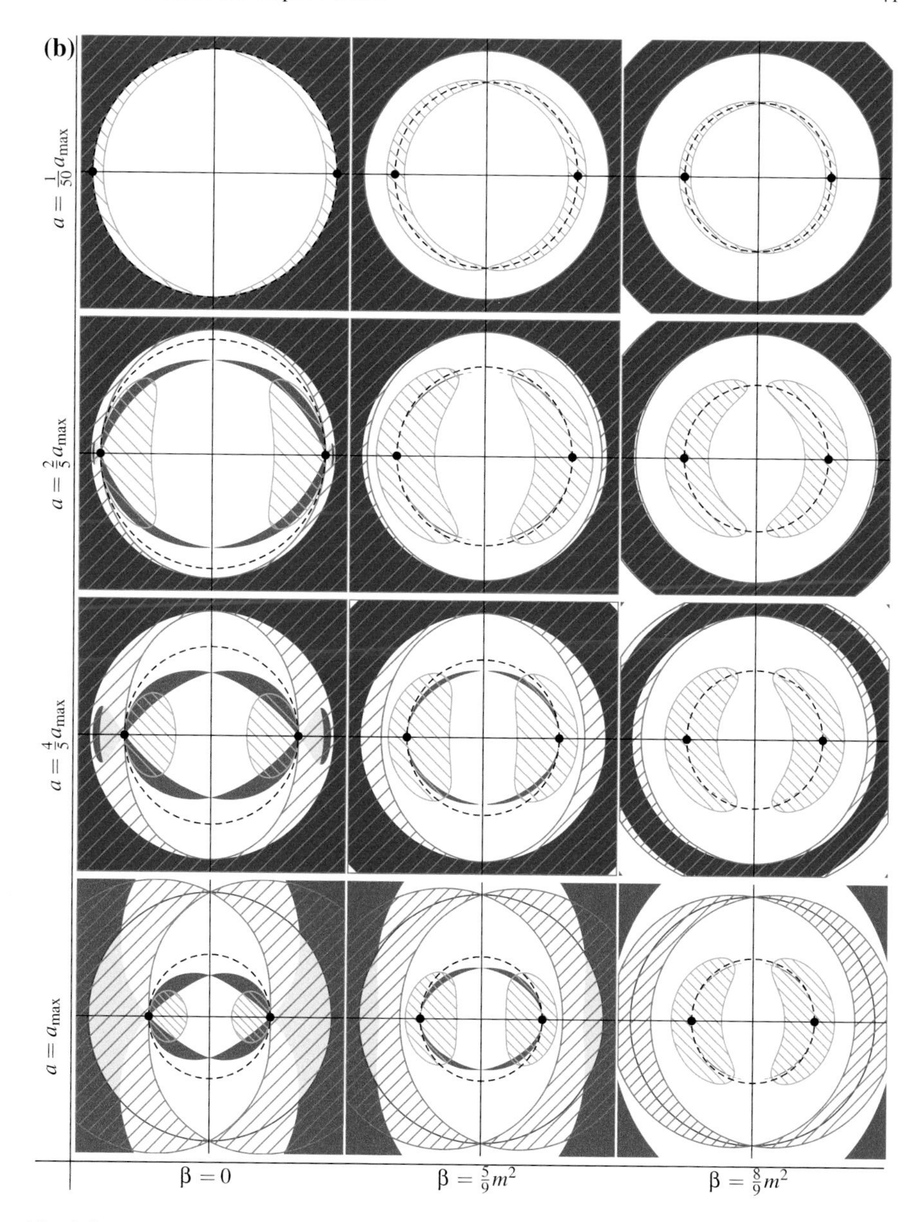

Fig. 3.4 (continued)

changes the symmetry behavior significantly, see Fig. 3.5. The plots are no longer symmetric with respect to the equatorial plane (but they remain, of course, axially symmetric). The exterior and interior photon regions and also the interior part of the causality violating region clearly show this asymmetry. The sign of ℓ determines the nature of the symmetry violation, see Fig. 3.7.

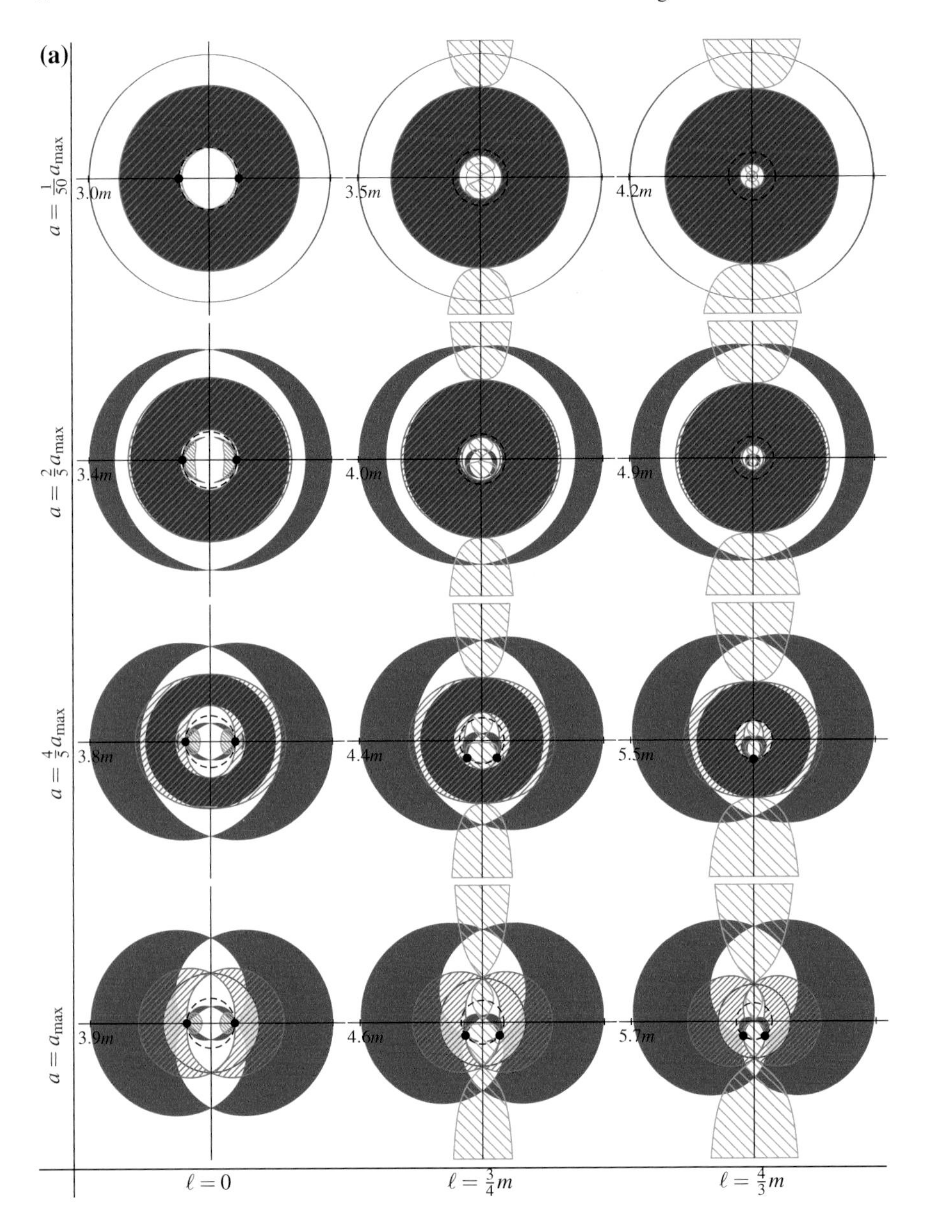

Fig. 3.5 Photon regions in Kerr–NUT space-times with $C = 0$ for spins $a = \lambda a_{\max}$ where $a_{\max} = \sqrt{m^2 + \ell^2}$. The *black hole* parameters are listed in Table A.2. **a** Photon regions in Kerr–NUT space-times ($C = 0$). **b** Magnified inner part of plots in (**a**)

For a slowly rotating Kerr–NUT black hole, $a^2 < \ell^2$, there is no ring singularity, and there are no stable spherical light rays. If the spin is increased, the ring singularity appears at $a^2 = \ell^2$, degenerated to a point on the rotational axis off the equatorial

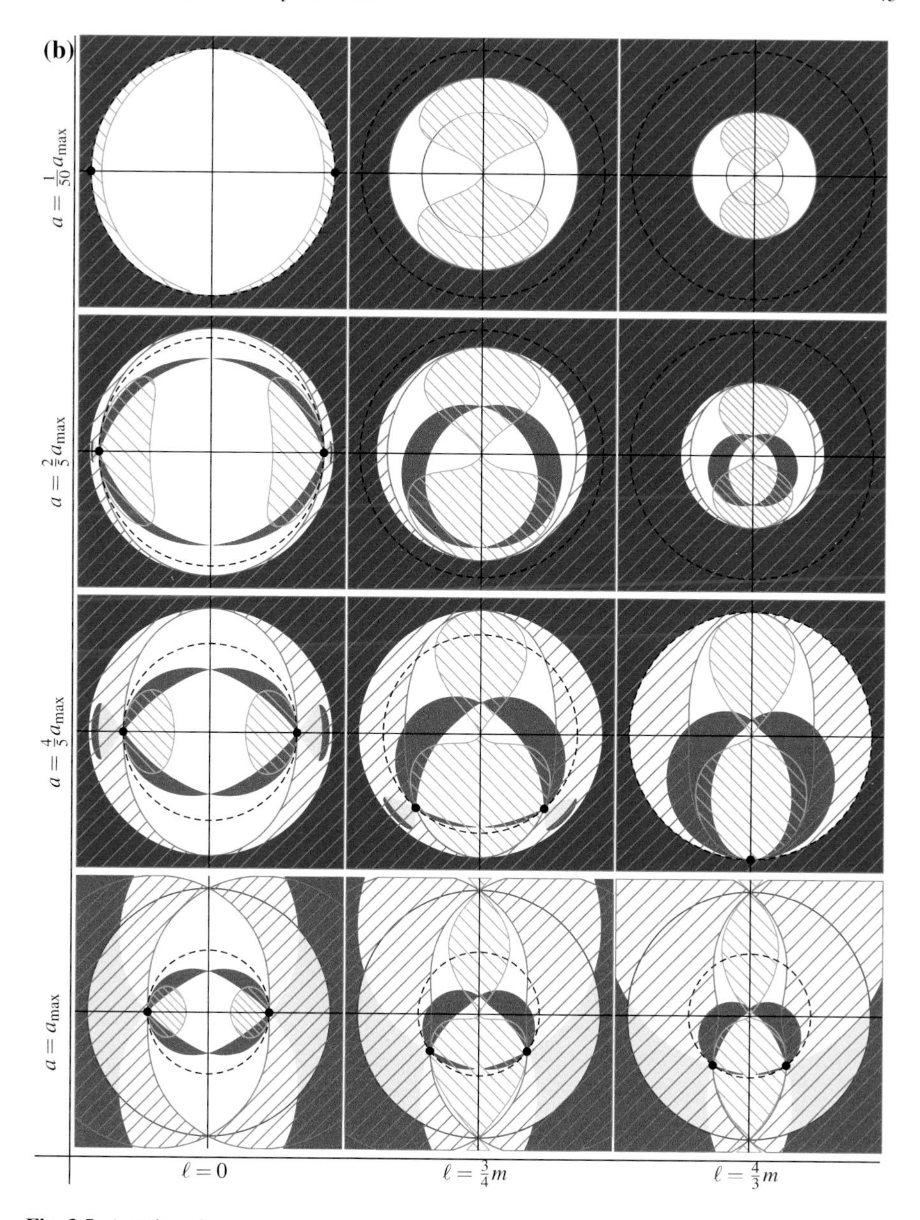

Fig. 3.5 (continued)

plane. With further increase of a, the ring singularity moves towards the equator and stable spherical light orbits appear between $r = 0$ and $r = r_-$; as in the Kerr case, the interior photon region consists of two connected components that are separated by the ring singularity. In contrast to the ergosphere, which stays symmetric

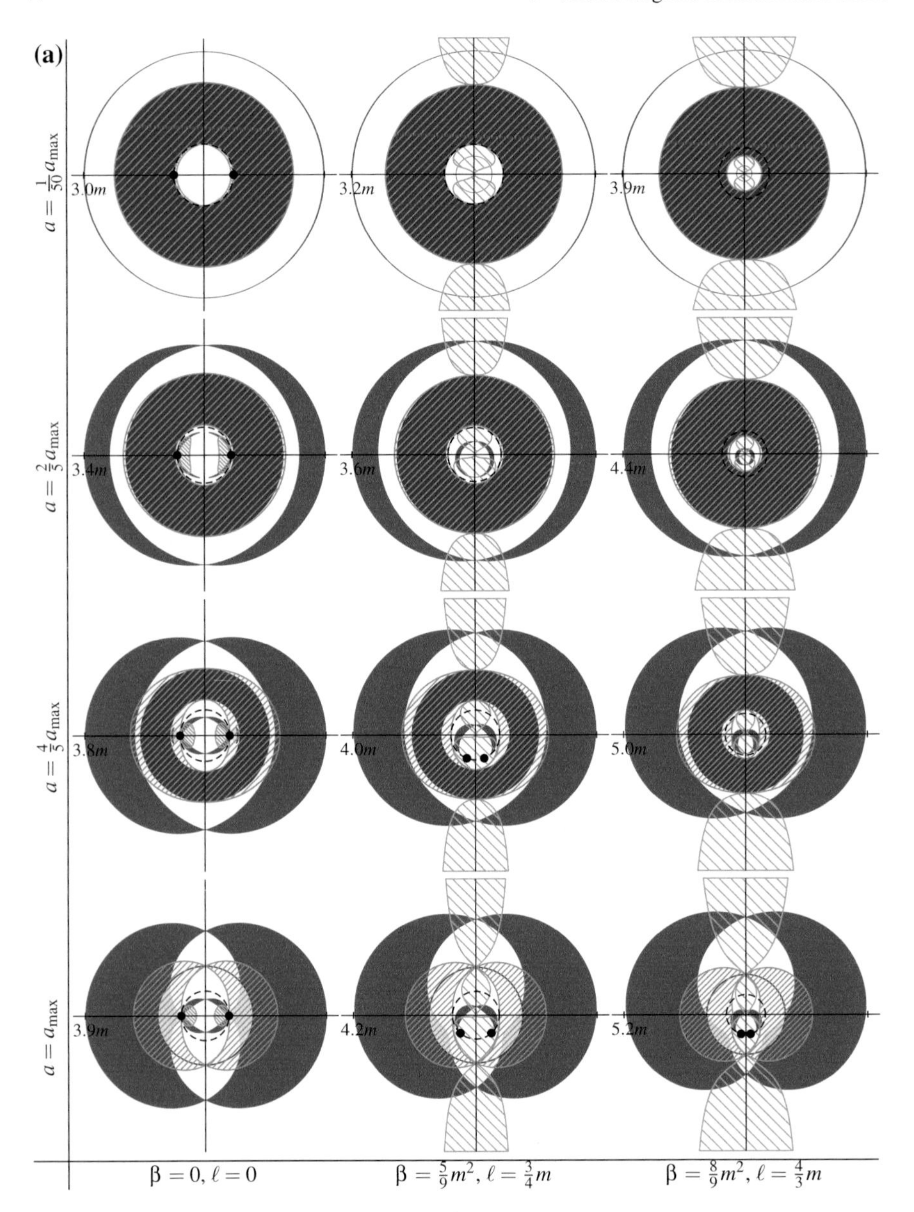

Fig. 3.6 Photon regions in Kerr–Newman–NUT space-times with $C = 0$ for spins $a = \lambda a_{\max}$ where $a_{\max} = \sqrt{m^2 + \ell^2 - \beta}$. The *black hole* parameters are listed in Table A.2. **a** Photon regions in Kerr–Newman–NUT space-times ($C = 0$). **b** Magnified inner part of plots in (**a**)

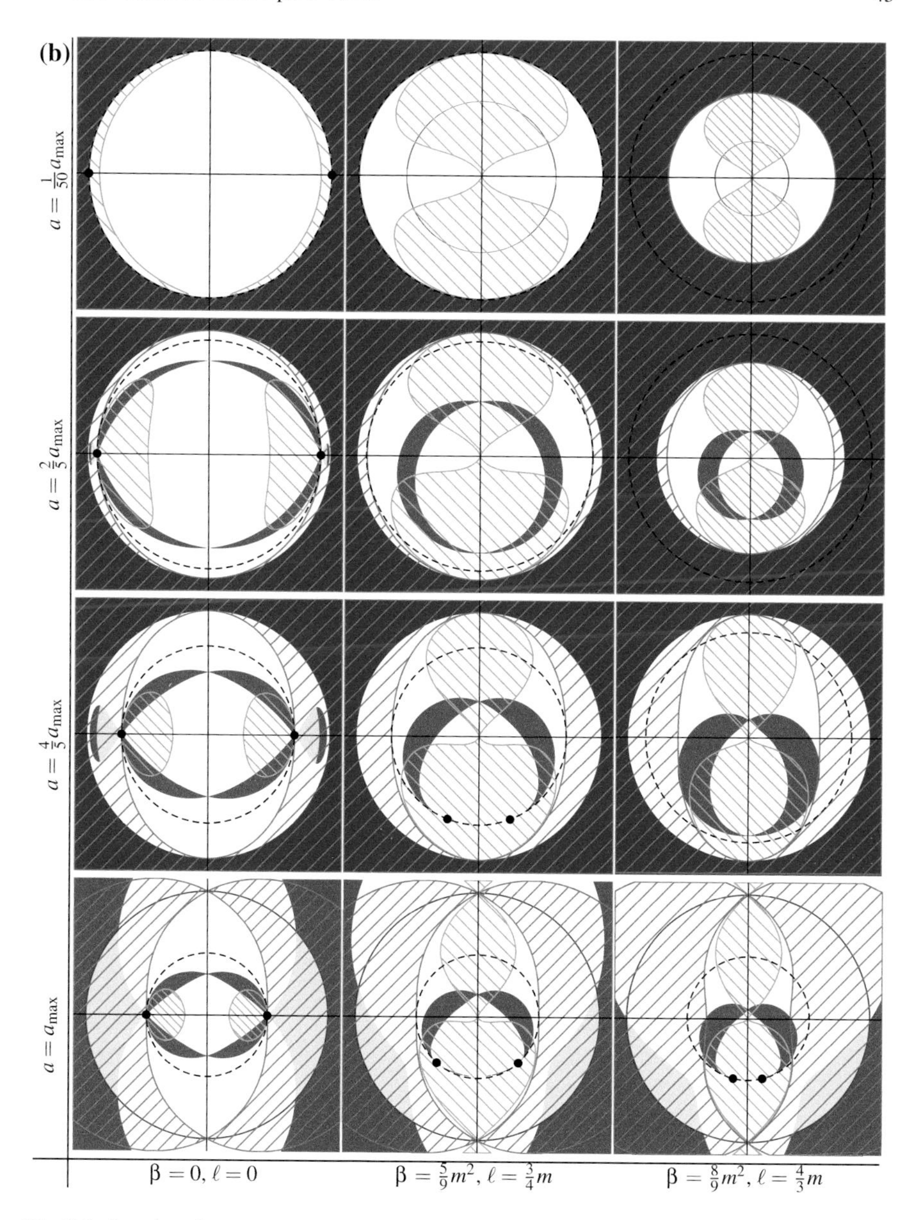

Fig. 3.6 (continued)

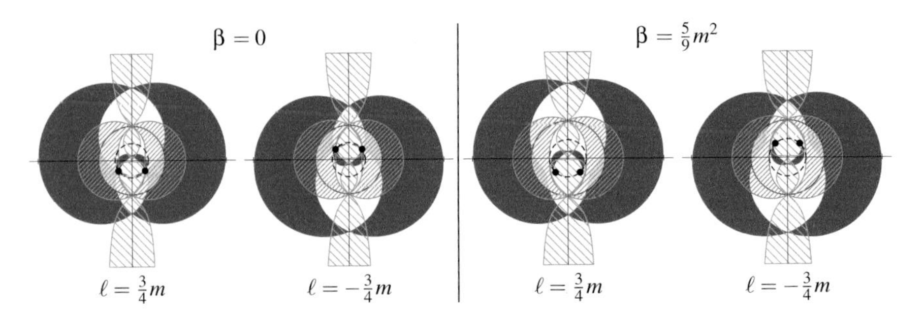

Fig. 3.7 Photon regions for different signs of the NUT parameter ℓ, where $C = 0$, $a = a_{\max} = \sqrt{m^2 + \ell^2 - \beta}$. By changing the sign of ℓ, the asymmetric regions are mirrored at the equatorial plane

(see Sect. 2.4) and is not significantly affected by ℓ, the causality violating region changes fundamentally. Now there is an additional causality violating region in the domain of outer communication; it is located around the z axis and encloses the axial singularity which is also caused by ℓ. This external region extends from the outer horizon at $r = r_+$ to $r = \infty$ while the interior causality violating region is now extending from the inner horizon at $r = r_-$ to $r = -\infty$ which corresponds to the origin in the extended polar plots. In addition to the changes mentioned before, the causality violating region depends similar to the axial singularity on the Manko–Ruiz parameter C which was chosen equal to zero in Fig. 3.5. The dependency on C is investigated in the more general Kerr–Newman–NUT space-time, see below and Fig. 3.8 for the appropriate graphics for other values of C.

Combinations of these phenomena are visible in the axially symmetric plots in Fig. 3.6 belonging to the **Kerr–Newman–NUT** space-time which contains the up to now considered space-times. As in the Kerr–NUT space-time, the plots are not symmetric to the equatorial plane in consequence of the gravitomagnetic charge ℓ. The sign of ℓ determines how the plots are deformed asymmetrically, cf. Fig. 3.7. The effects of the charge β are also preserved. Again, the two connected components of the interior photon region get detached from the ring singularity with a charge β but now if $a^2 > \ell^2$. And also highly charged fast rotating black holes are similar to slowly rotating but hardly charged black holes. Since the condition—Eq. (2.13)—for the existence of the ring singularity is independent of β, there are no changes in comparison to the Kerr–NUT space-time. Furthermore, the ergoregion stays symmetric and the causality violating region appears in the domain of outer communication, too.

As mentioned before, the causality violating region depends on the Manko–Ruiz parameter C. The appropriate images for fixed Kerr–Newman–NUT space-time with $C \in \{-2, -1, -\frac{1}{2}, 0, \frac{1}{2}, 1, 2\}$ are shown in Fig. 3.8. Since there are no changes except for the causality violating region, we added a cosmological constant $\Lambda = 10^{-2}\,\mathrm{m}^{-2}$ for positive Manko–Ruiz Parameter C in anticipation of the following Sect. 3.2. The

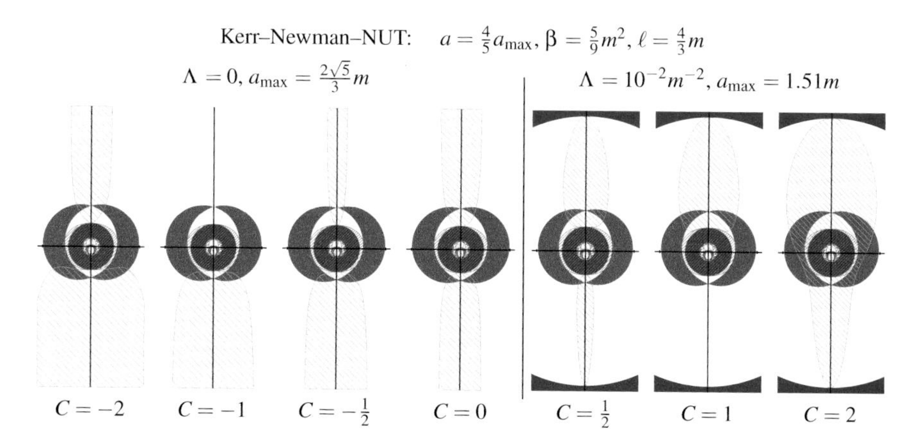

Fig. 3.8 Causality violation for varying singularity parameter C. If $C = \pm 1$, one of the two half-axes is regular and not surrounded by a causality violating region which is consistent with Fig. 2.2. For $\Lambda \neq 0$, the additional cosmological horizon restricts the region (hatched) where the causality is violated. Parameters are listed in Table A.3

corresponding plots for completing both series are shown in Fig. 6 in Grenzebach et al. (2014).

Compared to Fig. 2.2, we see that the exterior causality violating region is always situated along the singularity on the z axis. This is also the case for $C \neq \pm 1$ where one of the two semi-axes is regular. Here, the causality is not violated on the regular part. With changing C, the size of the causality violating region varies which goes along with the increasing/decreasing angular momenta of the semi-infinite rotating rods. Hence, the parameter C is not only balancing the distribution of the singularity but also of the causality violating region. Bigger values of C lead to the joining of the two disconnected causality violating regions, for instance at $C = 2$ in Fig. 2.2. Then they cover the entire z axis tubular.

3.2 Space-Times with Cosmological Constant

Here, we consider the Plebański class of black hole space-times which consists of Kerr–Newman–NUT space-times amended with the cosmological constant Λ, compare Table 2.1. Figure 3.9 comprises several plots of photon regions around maximally rotating black holes for those four subclasses we have discussed for $\Lambda = 0$ in the previous section. We only consider extremal black holes here since we are interested in the consequences of the cosmological constant; effects of the spin can be deduced from the previously discussed space-times. For the sake of completeness, Table A.5 contains also the parameters for slower rotating black holes. For the pictures, we have chosen a (small and) positive value for Λ such that the domain of outer communication is bounded by a cosmological horizon. The latter is not shown in

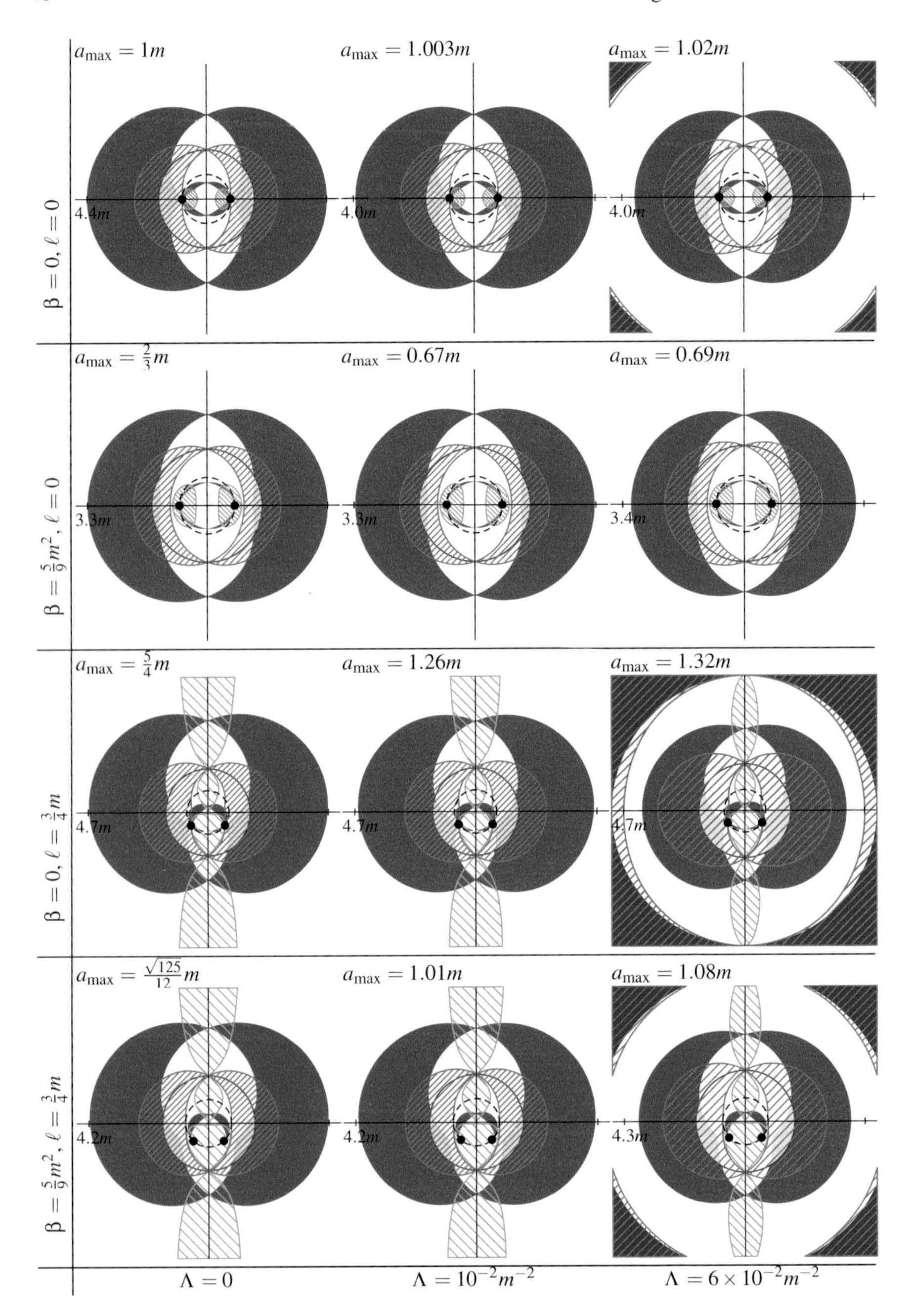

Fig. 3.9 Photon regions in Plebański space-times with cosmological constant Λ for a fixed extremal spin $a = a_{\max}$. Table A.5 contains a complete list of parameters

$\beta = \frac{5}{9}m^2, \quad \ell = \frac{3}{4}m, C = 0$
$\Lambda = 6 \times 10^{-2}m^{-2}$
$a_{\max} = 1.08m$

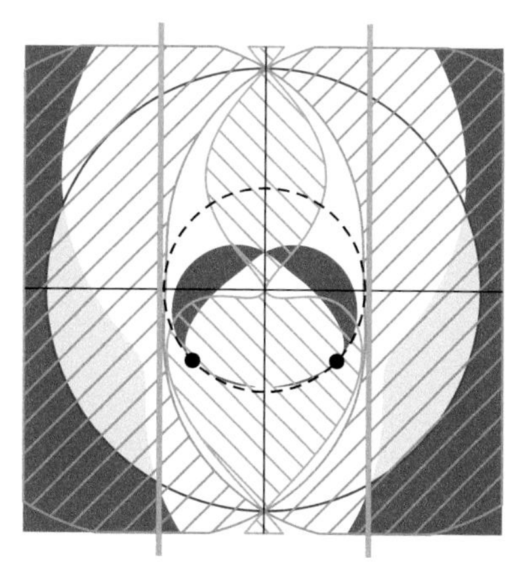

Fig. 3.10 Magnified inner part of an asymmetric ergoregion ▨ in Kerr–Newman–NUT space-time with cosmological constant Λ. Even with the *green auxiliary lines*, the asymmetry is hard to recognize

all pictures in Fig. 3.9 because these pictures do not extend so far. The cosmological horizon restricts the causality violating region which depends on the Manko–Ruiz parameter C, see Fig. 3.8, and it is covered by a second new part of the ergoregion.

Besides these changes, the cosmological constant Λ affects the shape or the size of the regions only marginally, see Fig. 3.9. The reason for this is of course the apparently small value of Λ, but compared to a realistic[4] value of Λ, the ones used for generating the images are already extremely large. Nevertheless, the biggest deviation besides the additional horizon is that the ergoregion is not symmetric any longer, see Eq. (2.24) with Eq. (2.7). But also this asymmetry is hardly recognizable. The best candidate would be the plot in the bottom right corner in Fig. 3.9 and actually the asymmetry is visible in the magnification of the inner part shown in Fig. 3.10.

Moreover, for non-zero Λ higher spin values $a_{\max}$ are possible compared to space-times with $\Lambda = 0$ where $a_{\max} = \sqrt{m^2 + \ell^2 - \beta}$, cf. Tables A.2 and A.5. But for $\Lambda \neq 0$ there is no convenient formula for $a_{\max}$ because one has to evaluate a fourth-order polynomial.

3.3 Accelerated Black Holes

The acceleration parameter is the last missing parameter of the Plebański–Demiański class for which we have not yet discussed its influence on the regions around a black hole. The appropriate images are contained in Fig. 3.11 where the subclass of the

[4]As realistic value of Λ one finds $\Lambda \approx +10^{-52}\,\mathrm{m}^{-2}$ (Unsöld and Baschek 2005) which corresponds to $\Lambda \approx 10^{-122}$ measured in Planck units (Barrow and Shaw 2011; Riess et al. 1998; Perlmutter et al. 1999).

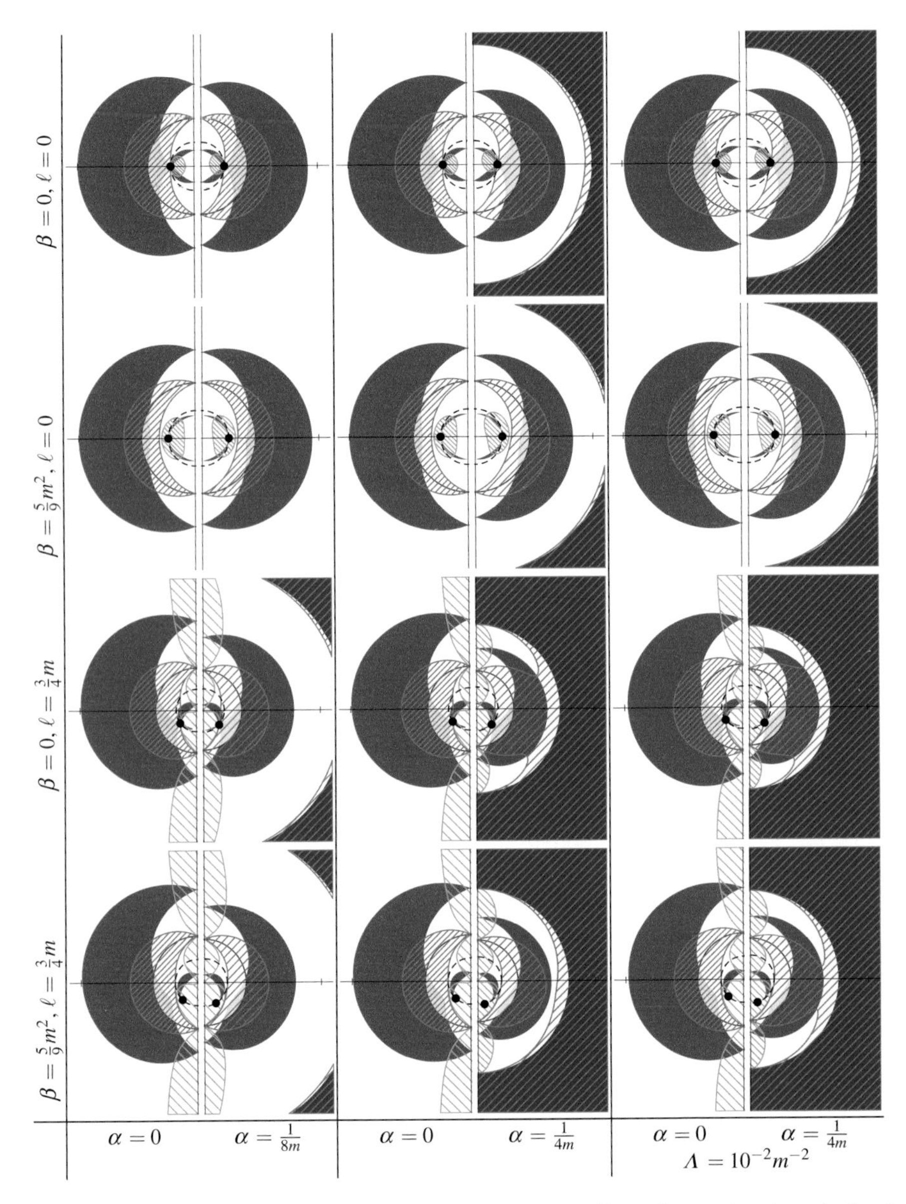

Fig. 3.11 Photon regions in Plebański–Demiański space-times with varying acceleration α for fixed spin $a = a_{\max}$, see Table A.6 for values of $a_{\max}$. In each column, the plots for the unaccelerated cases (*left*) are opposed to the accelerated ones (*right*)

space-times is varied from line to line as in Fig. 3.9. Because the consequences of the spin were discussed in detail earlier, we consider again only maximally rotating black holes but list the space-time parameters also for slower spinning black holes, see Table A.6. Pictures for slower rotating accelerated Kerr black holes can be found in

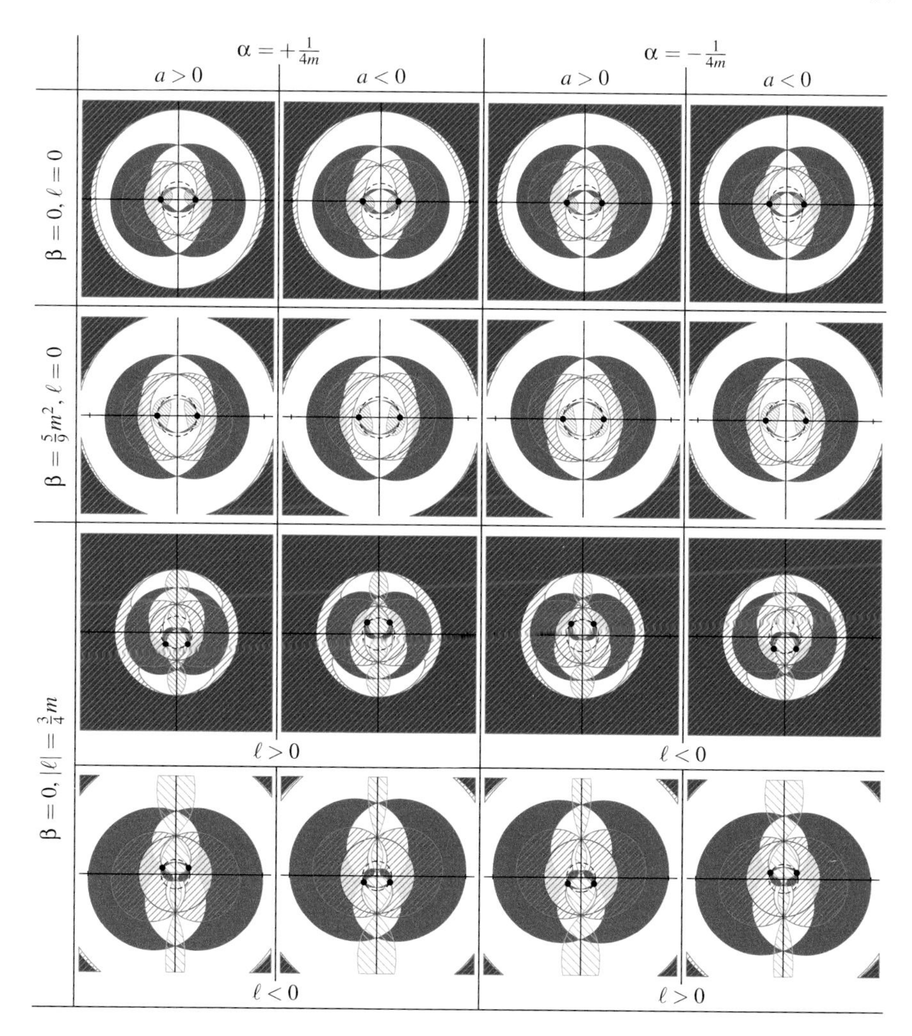

Fig. 3.12 Photon regions for different signs of the spin a, the NUT charge ℓ ($C = 0$) and the acceleration α for extremal Kerr, Kerr–Newman and Kerr–NUT space-times. Values for $|a| = a_{\max} = \sqrt{m^2 + \ell^2 - \beta}$ are noted in Table A.7. The asymmetric regions affected by the change of a sign get mirrored at the equatorial plane

Grenzebach et al. (2015) in Fig. 2. Even though in reality only very small acceleration parameters are expected, we choose relatively large values ($\alpha \in \left\{0, \frac{1}{8m}, \frac{1}{4m}\right\}$) for a better illustration of the effects, similar as for the cosmological constant.

The plots for the accelerated space-times look similar to the non-accelerated ones. There are no new qualitative effects of β and Λ if α is present. But there are two significant differences. Firstly, a non-zero acceleration parameter gives rise to additional horizons, analogous to a cosmological constant. Secondly, the plots are no longer symmetric with respect to the equatorial plane which is similar to the

Table 3.1 Sign changes for which regions get mirrored at the equatorial plane

	Kerr	Kerr–Newman	Kerr–NUT
Photon region $\mathscr{K}$ (▇, ▇)	$\alpha \mapsto -\alpha$	$\alpha \mapsto -\alpha$	$a\alpha \mapsto -a\alpha$
Causality violation (▧)	$a\alpha \mapsto -a\alpha$	$a\alpha \mapsto -a\alpha$	$a\alpha \mapsto -a\alpha$
Ergoregion (▨)	$a\alpha \mapsto -a\alpha$	$a\alpha \mapsto -a\alpha$	$a\alpha \mapsto -a\alpha$

NUT case. The additional outer horizon, a cosmological one, is seen in almost all illustrations. In principle, such a horizon appears in all plots but it happens that most or even all of it is located outside of the shown clipping. Only for slowly rotating black holes (not shown here) with dominant NUT property and $\ell\alpha < 0$ there is no additional cosmological horizon that restricts the domain of outer communication. Then, Δ_r has three negative roots as explained in the paragraph about the horizons in Sect. 2.3; see Table A.7 for examples. The asymmetry with respect to the equatorial plane is best seen for $\alpha = \frac{1}{4m}$. With the exception of the causality violating region, the entire picture looks as if pushed into the negative z direction, i.e., into the direction against the direction of the acceleration.

As expected, the photon region, the ergoregion and the causality violating region (more precisely, their deformations) depend on the signs of a, ℓ and α. From Fig. 3.12 we see that these regions are mirrored at the equatorial plane if the sign of α alone or of the product $a\alpha$ is changed, see Table 3.1.

Considering the plots in Fig. 3.12 for the Kerr–NUT space-time, one notices that these are in fact very different from line to line but very similar within a line. This behavior is in accord with the isometries discussed in Sect. 2.2 since ℓ and α have opposite signs in the last row while the signs coincide in the row next to the last.

3.4 Summary

In all considered space-times, the photon region $\mathscr{K}$ develops a crescent-shaped cross-section $\mathfrak{D}$ if the described black hole rotates. For the Kerr, Kerr–Newman (KN), Kerr–NUT (KNUT) and Kerr–Newman–NUT (KNNUT) space-times, we summarize in Table 3.2 whether $a_{\max}$ is increased or decreased in comparison to Kerr space-time.

Table 3.2 Summary of properties of different space-times

Space-time	$a_{\max}$	Photon region	Ring singularity	Causality violation
Kerr		Spherical symm.	Exist	Hidden by horizon
KN	Decreased	Spherical symm.	Exist	Hidden by horizon
KNUT	Increased	Rotational symm.	Exists if $a^2 \geq \ell^2$	In dom. of out. com.
KNNUT	Increased	Rotational symm.	Exists if $a^2 \geq \ell^2$	In dom. of out. com.

Furthermore, we indicate the symmetry property of the photon region $\mathcal{K}$, the existence of a ring singularity and where the causality gets violated. The ergoregion is always symmetric unless $\Lambda \neq 0$ or $\alpha \neq 0$. If $\Lambda \neq 0$ or $\alpha \neq 0$, then there exists an additional cosmological horizon in the domain of outer communication.

By [1–3] I refer to my papers Grenzebach et al. (2014), Grenzebach (2015) and Grenzebach et al. (2015), respectively. Sentences marked with [i] can be found in total or only slightly modified in the ith paper

References

Barrow JD, Shaw DJ (2011) The value of the cosmological constant. Gen Relativ Gravit 43(10):2555–2560. doi:10.1007/s10714-011-1199-1

Carter B (1968) Hamilton-Jacobi and Schrödinger separable solutions of Einstein's equations. Commun Math Phys 10(4):280–310. http://projecteuclid.org/euclid.cmp/1103841118

Grenzebach A, Perlick V, Lämmerzahl C (2014) Photon regions and shadows of Kerr–Newman–NUT Black Holes with a cosmological constant. Phys Rev D 89:124,004(12). doi:10.1103/PhysRevD.89.124004. arXiv:1403.5234

Grenzebach A (2015) Aberrational effects for shadows of black holes. In: Puetzfeld et al Proceedings of the 524th WE-Heraeus-Seminar "Equations of Motion in Relativistic Gravity", held in Bad Honnef, Germany, 17–23 Feb 2013, pp 823–832. doi:10.1007/978-3-319-18335-0_25, arXiv:1502.02861

Grenzebach A, Perlick V, Lämmerzahl C (2015) Photon Regions and shadows of accelerated black holes. Int J Mod Phys D 24(9):1542,024(22). doi:10.1142/S0218271815420249 (Special Issue "Papers" of the "7th Black Holes Workshop", Aveiro, Portugal, arXiv:1503.03036)

Hackmann E (2010) Geodesic equations in black hole space-times with cosmological constant. Dissertation, Universität Bremen, Bremen. http://nbn-resolving.de/urn:nbn:de:gbv:46-diss000118806

Hackmann E, Kagramanova V, Kunz J, Lämmerzahl C (2009) Analytic solutions of the geodesic equation in axially symmetric space-times. Europhys Lett 88(3):30,008(5). doi:10.1209/0295-5075/88/30008

Hartle JB (2003) Gravity: An Introduction to Einstein's General Relativity. Pearson Education (Addison-Wesley), San Francisco

Kagramanova V, Kunz J, Hackmann E, Lämmerzahl C (2010) Analytic treatment of complete and incomplete geodesics in Taub–NUT space-times. Phys Rev D 81(12):124,044(17). doi:10.1103/PhysRevD.81.124044

Perlick V (2004) Gravitational lensing form a spacetime perspective. Living Rev Relativ 7(9). doi:10.12942/lrr-2004-9

Perlmutter S, Aldering G, Goldhaber G, Knop RA, Nugent P, Castro PG, Deustua S, Fabbro S, Goobar A, Groom DE, Hook IM, Kim AG, Kim MY, Lee JC, Nunes NJ, Pain R, Pennypacker CR, Quimby R, Lidman C, Ellis RS, Irwin M, McMahon RG, Ruiz-Lapuente P, Walton N, Schaefer B, Boyle BJ, Filippenko AV, Matheson T, Fruchter AS, Panagia N, Newberg HJM, Couch WJ, Project TSC (1999) Measurements of Ω and Λ from 42 High-Redshift Supernovae. Astrophys J 517(2):565. doi:10.1086/307221

Riess AG, Filippenko AV, Challis P, Clocchiatti A, Diercks A, Garnavich PM, Gilliland RL, Hogan CJ, Jha S, Kirshner RP, Leibundgut B, Phillips MM, Reiss D, Schmidt BP, Schommer RA, Smith RC, Spyromilio J, Stubbs C, Suntzeff NB, Tonry J (1998) Observational evidence from supernovae for an accelerating universe and a cosmological constant. Astron J 116(3):1009. doi:10.1086/300499

Teo E (2003) Spherical Photon Orbits around a Kerr Black Hole. Gen Relativ Gravit 35(11):1909–1926. doi:10.1023/A:1026286607562

Unsöld A, Baschek B (2005) Der neue Kosmos, 7th edn. Springer, Heidelberg

Chapter 4
The Shadow of Black Holes

Abstract It is explained how to derive analytical formulas for the boundary curve of the shadow as seen by an observer at given position in the domain of outer communication. The formulas are used to analyze the dependency of the shadow of a black hole on the motion of the observer. Furthermore, the horizontal and vertical angular diameters of the shadow are calculated. Although explicit formulas are given for the Kerr space-time only, the method holds true for the general Plebański–Demiański class. After all, the angular diameters for the black holes at the centers of our Galaxy and of M87 are estimated.

Keywords Shadow black hole · Shadow analytic formula · Shadow boundary curve · Moving observer · Inclination observer · Galactic black hole · Penrose aberration · Angular diameter shadow · Sgr A* shadow · M87 shadow

For an observer pointing the telescope into the direction of a black hole, there is a region on the sky which stays dark, provided that there are no light sources between the observer and the black hole.[1] This dark region is called the *shadow* of the black hole. To determine the shape of the shadow, it is convenient to consider light rays which are sent *into the past* from the position[2] (r_O, ϑ_O) of a fixed observer in the domain of outer communication. Then we can distinguish between two types of lightlike geodesics: Those where the radial coordinate increases to infinity after possibly passing through a minimum and those where the radial coordinate decreases until reaching the horizon at $r = r_+$. If we assume that there is a distribution of light sources in the universe, excluding the region between the observer and the black hole, geodesics of the first kind could reach a light source; so we assign brightness to the initial direction of such a light ray. Vice versa, we assign darkness to the initial directions of light rays of the second kind, i.e., these initial directions determine the shadow of the black hole. The boundary of the shadow corresponds to light rays on the borderline between the two kinds. These light rays spiral asymptotically towards one of the unstable spherical light orbits in the exterior photon region $\mathscr{K}$ as discussed

[1] Parts of this section are taken from my three papers. The Sects. 4.1, 4.5 are based on [1], Sect. 4.4 on [2] and Sects. 4.3, 4.6 on [3].

[2] Because of the symmetry of the Plebański–Demiański space-time, it is enough to specify the r and ϑ coordinate to define a fixed position in space-time.

A. Grenzebach, *The Shadow of Black Holes*,
SpringerBriefs in Physics, DOI 10.1007/978-3-319-30066-5_4

in the Chap. 3. Hence, the essential information for determining the shadow of a black hole is in the surrounding photon region $\mathcal{K}$ given by (3.7). One may even say that the shadow is an image of the photon region (but not of the event horizon).

In the following, we derive analytical formulas for the boundary curve of the shadow seen by our fixed observer and by observers at the same position (r_O, ϑ_O) that are moving relatively against the fixed observer. At first, we choose an orthonormal tetrad (Griffiths and Podolský 2009, p. 307) for the fixed observer, see Fig. 4.1

$$e_0 = \Omega \left. \frac{a\partial_\varphi + (\Sigma + a\chi)\partial_t}{\sqrt{\Sigma \Delta_r}} \right|_{(r_O,\vartheta_O)}, \qquad e_2 = -\Omega \left. \frac{\partial_\varphi + \chi\partial_t}{\sqrt{\Sigma \Delta_\vartheta}\sin\vartheta} \right|_{(r_O,\vartheta_O)},$$
$$e_1 = \Omega \left. \sqrt{\frac{\Delta_\vartheta}{\Sigma}}\,\partial_\vartheta \right|_{(r_O,\vartheta_O)}, \qquad e_3 = -\Omega \left. \sqrt{\frac{\Delta_r}{\Sigma}}\,\partial_r \right|_{(r_O,\vartheta_O)}. \tag{4.1}$$

It is chosen such that $e_0 \pm e_3$ are aligned with both *principal null congruences* of our metric. The basis vector e_0 is interpreted as the four-velocity of the observer because it is a timelike vector; e_3 points into the spatial direction towards the center of the black hole. An observer with this tetrad is called a *standard observer* in the following.

Observe that Δ_r is positive since the standard observer is in the domain of outer communication. Moreover, Σ is positive everywhere (except at the ring singularity which is not part of the space-time and, moreover, away from the domain of outer communication) and Δ_ϑ is positive by assumption, see Sect. 2.3. This setting guarantees real coefficients in Eq. (4.1) and it is easy to check that the e_i are indeed orthonormal.

Of course, the shape of the shadow depends on the observer's state of motion. Therefore, we have to modify the chosen tetrad (4.1) if another observer at (r_O, ϑ_O) moves with velocity $\mathbf{v} = (v_1, v_2, v_3)$, $v = |\mathbf{v}| < 1 = c$, relative to our standard observer. The four-velocity of the moving observer is

$$\widetilde{e}_0 = \frac{v_1 e_1 + v_2 e_2 + v_3 e_3 + e_0}{\sqrt{1 - v^2}}. \tag{4.2a}$$

From $\{\widetilde{e}_0, e_1, e_2, e_3\}$ we find an orthonormal tetrad $\{\widetilde{e}_0, \widetilde{e}_1, \widetilde{e}_2, \widetilde{e}_3\}$ with the Gram–Schmidt procedure[3] by adding e_3, e_1, e_2—in this order—successively to $\widetilde{e}_0$

[3] Gram–Schmidt orthonormalization: $\widetilde{e}_3 \propto e_3 + g(e_3, \widetilde{e}_0)\widetilde{e}_0$, $\widetilde{e}_1 \propto e_1 + g(e_1, \widetilde{e}_0)\widetilde{e}_0 - g(e_1, \widetilde{e}_3)\widetilde{e}_3$, $\widetilde{e}_2 \propto e_2 + g(e_2, \widetilde{e}_0)\widetilde{e}_0 - g(e_2, \widetilde{e}_1)\widetilde{e}_1 - g(e_2, \widetilde{e}_3)\widetilde{e}_3$.

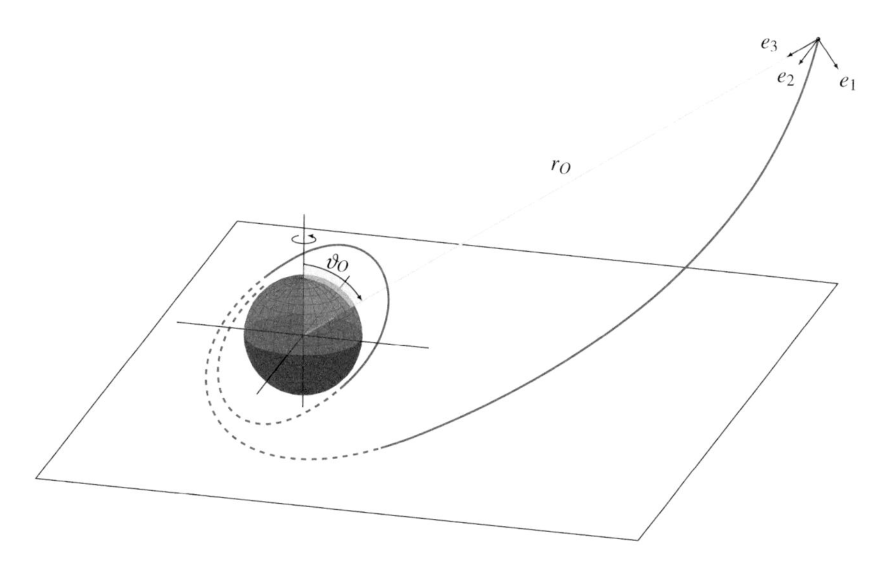

Fig. 4.1 (After [1]) At an observation event with Boyer–Lindquist coordinates (r_O, ϑ_O) we choose an orthonormal tetrad $\{e_0, e_1, e_2, e_3\}$ according to Eq. (4.1)

$$\begin{aligned}
\widetilde{e}_1 &= \frac{\big(1 - v_2^2\big) e_1 + v_1 (v_2 e_2 + e_0)}{\sqrt{1 - v_2^2}\,\sqrt{1 - v_1^2 - v_2^2}}, \\
\widetilde{e}_2 &= \frac{e_2 + v_2 e_0}{\sqrt{1 - v_2^2}}, \\
\widetilde{e}_3 &= \frac{\big(1 - v_1^2 - v_2^2\big) e_3 + v_3 (v_1 e_1 + v_2 e_2 + e_0)}{\sqrt{1 - v_1^2 - v_2^2}\,\sqrt{1 - v^2}}.
\end{aligned} \tag{4.2b}$$

Note that $\widetilde{e}_i = e_i$ if $v_i = 0$, i.e., for $\mathbf{v} = 0$ this procedure recovers the tetrad $\{e_0, e_1, e_2, e_3\}$ from (4.1). As before, the spacelike vector $\widetilde{e}_3$ corresponds to the direction towards the black hole. The interpretation of $\widetilde{e}_1$ and $\widetilde{e}_2$ becomes clear if we introduce celestial coordinates, see Eq. (4.4) and Fig. 4.2. Then, $\widetilde{e}_1$ and $\widetilde{e}_2$ point into the North–South and the West–East direction, respectively.

For any light ray $\lambda(s) = \big(r(s), \vartheta(s), \varphi(s), t(s)\big)$, the tangent vector at the position of the observer can be written in two different ways, using either the Boyer–Lindquist coordinate basis or the tetrad (4.2a, 4.2b) introduced above

$$\dot{\lambda} = \dot{r}\partial_r + \dot{\vartheta}\partial_\vartheta + \dot{\varphi}\partial_\varphi + \dot{t}\partial_t, \tag{4.3}$$

$$\dot{\lambda} = \sigma\big(-\widetilde{e}_0 + \sin\theta\cos\psi\;\widetilde{e}_1 + \sin\theta\sin\psi\;\widetilde{e}_2 + \cos\theta\;\widetilde{e}_3\big). \tag{4.4}$$

The second equation defines the celestial coordinates θ and ψ, see Fig. 4.2, where $\theta = 0$ corresponds to the direction towards the black hole. The scalar factor σ can be

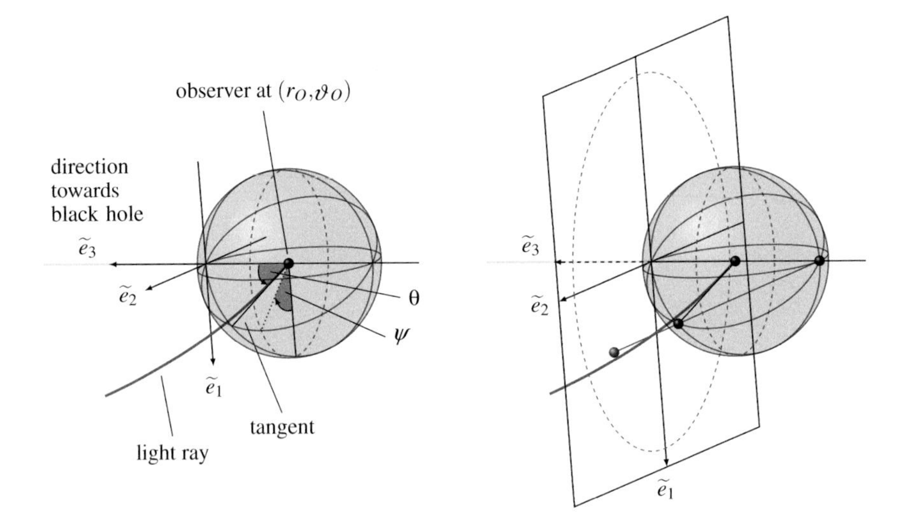

Fig. 4.2 (After [2]) *On the left*: The direction of each light ray reaching the observer at (r_O, ϑ_O) is given by the celestial coordinates θ and ψ (Eq. 4.4) of their tangents. *On the right*: The points (θ, ψ) on the celestial sphere (*black* ball) can be identified with points in the plane (*red* ball) by stereographic projection. The *dashed red circles* mark the celestial equator $\theta = \pi/2$ and its projection

calculated from $\widetilde{e}_0$ since $\sigma = g\big(\dot{\lambda}, \widetilde{e}_0\big)$, see (4.10) below. For the tetrad (4.2a, 4.2b), we observe the following dependencies

$$\begin{aligned}
\widetilde{e}_0 &= k_{0r}\partial_r + k_{0\vartheta}\partial_\vartheta + k_{0\varphi}\partial_\varphi + k_{0t}\partial_t, \\
\widetilde{e}_1 &= k_{1\vartheta}\partial_\vartheta + k_{1\varphi}\partial_\varphi + k_{1t}\partial_t, \\
\widetilde{e}_2 &= k_{2\varphi}\partial_\varphi + k_{2t}\partial_t, \\
\widetilde{e}_3 &= k_{3r}\partial_r + k_{3\vartheta}\partial_\vartheta + k_{3\varphi}\partial_\varphi + k_{3t}\partial_t
\end{aligned} \tag{4.5}$$

where the $k_{\mu\nu}$ are the coefficients of the basis $\widetilde{e}_\mu$ regarding to the coordinate basis ∂_ν which can be read off Eq. (4.1). Hence

$$\begin{aligned}
\dot{\lambda} = \sigma\big(&(-k_{0r} + k_{3r}\cos\theta)\partial_r + (-k_{0\vartheta} + k_{1\vartheta}\sin\theta\cos\psi + k_{3\vartheta}\cos\theta)\partial_\vartheta \\
&+ (-k_{0\varphi} + k_{1\varphi}\sin\theta\cos\psi + k_{2\varphi}\sin\theta\sin\psi + k_{3\varphi}\cos\theta)\partial_\varphi \\
&+ (-k_{0t} + k_{1t}\sin\theta\cos\psi + k_{2t}\sin\theta\sin\psi + k_{3t}\cos\theta)\partial_t\big).
\end{aligned} \tag{4.6}$$

Comparing coefficients of ∂_r, ∂_ϑ, and ∂_φ in (4.3) and (4.6) yields

$$\dot{r} = \sigma(-k_{0r} + k_{3r}\cos\theta), \tag{4.7a}$$

$$\dot{\vartheta} = \sigma(-k_{0\vartheta} + k_{1\vartheta}\sin\theta\cos\psi + k_{3\vartheta}\cos\theta), \tag{4.7b}$$

$$\dot{\varphi} = \sigma(-k_{0\varphi} + k_{1\varphi}\sin\theta\cos\psi + k_{2\varphi}\sin\theta\sin\psi + k_{3\varphi}\cos\theta). \tag{4.7c}$$

These equations can be solved easily for $\cos\theta$ and $\sin\psi$ (using $k_{1\vartheta}\sin\theta\cos\psi = \frac{1}{\sigma}\dot{\vartheta} + k_{0\vartheta} - k_{3\vartheta}\cos\theta$),

$$\cos\theta = \left.\frac{\frac{1}{\sigma}\dot{r} + k_{0r}}{k_{3r}}\right|_{(r_O,\vartheta_O)}, \tag{4.8a}$$

$$\sin\psi = \left.\frac{k_{3r}\left(\frac{1}{\sigma}\dot{\varphi} + k_{0\varphi} - \frac{k_{1\varphi}}{k_{1\vartheta}}\left(\frac{1}{\sigma}\dot{\vartheta} + k_{0\vartheta}\right)\right) - \left(k_{3\varphi} - \frac{k_{3\vartheta}}{k_{1\vartheta}}k_{1\varphi}\right)\left(\frac{1}{\sigma}\dot{r} + k_{0r}\right)}{k_{2\varphi}\sqrt{k_{3r}^2 - \left(\frac{1}{\sigma}\dot{r} + k_{0r}\right)^2}}\right|_{(r_O,\vartheta_O)}, \tag{4.8b}$$

where $\dot{\varphi}$, $\dot{\vartheta}$, and $\dot{r}$ have to be substituted from the equations of motion (3.3); since $\dot{r}$ and $\dot{\vartheta}$ are given as quadratic expressions, the signs have to be chosen consistently. More precisely, since we consider light rays that reach the observer, $\dot{r}$ occurs only with positive sign $\dot{r} = +\sqrt{\ldots}$ and because these light rays are parametrized in past direction, those with positive sign $\dot{\vartheta} = +\sqrt{\ldots}$ belong to the upper part of the shadow and those with negative sign $\dot{\vartheta} = -\sqrt{\ldots}$ to the lower part.

For calculating the remaining scalar factor σ, we express $\widetilde{e}_0$, Eq. (4.2a), in terms of the tetrad $\{\partial_r, \partial_\vartheta, \partial_\varphi, \partial_t\}$, Eq. (4.1)

$$\widetilde{e}_0 = \frac{\Omega}{\sqrt{\Sigma}\sqrt{1-v^2}}\left(\frac{(\Sigma + a\chi)\partial_t + a\partial_\varphi}{\sqrt{\Delta_r}} + v_1\sqrt{\Delta_\vartheta}\,\partial_\vartheta - v_2\frac{\partial_\varphi + \chi\,\partial_t}{\sqrt{\Delta_\vartheta}\sin\vartheta} - v_3\sqrt{\Delta_r}\,\partial_r\right). \tag{4.9}$$

As $\sigma = g(\dot{\lambda}, \widetilde{e}_0)$, see Eq. (4.4), we get σ from (2.1), (4.3), and (4.9)

$$\sigma = \frac{\Omega}{\sqrt{\Sigma}\sqrt{1-v^2}}\left(\frac{aL_z - (\Sigma + a\chi)E}{\sqrt{\Delta_r}} + \frac{v_1}{\sqrt{\Delta_\vartheta}}\frac{\Sigma}{\Omega^2}\dot{\vartheta} - v_2\frac{L_z - \chi E}{\sqrt{\Delta_\vartheta}\sin\vartheta} - \frac{v_3}{\sqrt{\Delta_r}}\frac{\Sigma}{\Omega^2}\dot{r}\right)\Bigg|_{(r_O,\vartheta_O)} \tag{4.10}$$

where $\dot{\vartheta}$ and $\dot{r}$ are given by (3.3) with the same sign rules as above.

For the standard observer with $\mathbf{v} = 0$, the coefficients $k_{\mu\nu}$ in Eq. (4.5) can be read off (4.1) since $\widetilde{e}_i = e_i$. With (4.10), Eq. (4.8a, 4.8b) simplifies to

$$\cos\theta = \left. \frac{\frac{\Sigma}{\Omega^2}\dot{r}}{(\Sigma + a\chi)E - aL_z} \right|_{(r_O,\vartheta_O)}, \tag{4.11a}$$

$$\sin\psi = \left. \frac{\sqrt{\Delta_\vartheta}\sin\vartheta}{\sqrt{\Delta_r}\sin\theta} \left(\frac{\frac{\Sigma}{\Omega^2}\dot{\varphi}\,\Delta_r}{(\Sigma + a\chi)E - aL_z} - a \right) \right|_{(r_O,\vartheta_O)}. \tag{4.11b}$$

One gets the same equations by comparing coefficients of ∂_φ and ∂_r in (4.3) and (4.4) with inserted e_i from (4.1). Since (4.11a, 4.11b) is much simpler than (4.8a, 4.8b), it is possible to write down the explicit expressions. Upon substituting for $\dot{\varphi}$ and $\dot{r}$ from (3.3), we find from (4.11a, 4.11b) that

$$T := \sin\theta = \left. \frac{\sqrt{\Delta_r K_E}}{r^2 + \ell^2 - a\widetilde{L}_E} \right|_{r_O}, \tag{4.12a}$$

$$P := \sin\psi = \left. \frac{\widetilde{L}_E + a\cos^2\vartheta + 2\ell\cos\vartheta}{\sqrt{\Delta_\vartheta K_E}\sin\vartheta} \right|_{\vartheta_O}, \tag{4.12b}$$

where

$$\widetilde{L}_E = L_E - a + 2\ell C. \tag{4.13}$$

Note that we do not have to take care of the signs since we substitute $\dot{\varphi}$ and $\dot{r}$ only but not $\dot{\vartheta}$.

We observe that the shadow is always symmetric with respect to a horizontal axis. The latter result follows from the fact that the points (ψ, θ) and $(\pi - \psi, \theta)$ correspond to the same constants of motion K_E and $\widetilde{L}_E$. For $\ell \neq 0$ and $\vartheta_O \neq \pi/2$ this symmetry property was not to be expected.

The Eqs. (4.8a, 4.8b) and (4.12a, 4.12b) are analytical parameter representations of the boundary curve of the black hole's shadow for a moving observer and the standard observer, respectively. The boundary represents lightlike geodesics which, if you think of sending them from the observer's position into the past, reach the photon region asymptotically. Each such geodesic must have the same constants of motion as the limiting spherical lightlike geodesic with radius r_p. By (3.6), the constants of motion of those light rays that correspond to boundary points of the shadow are given by

$$K_E = \left. \frac{16r^2\Delta_r}{(\Delta_r')^2} \right|_{r_p}, \qquad aL_E = \left(\Sigma + a\chi\right) - \left. \frac{4r\Delta_r}{\Delta_r'} \right|_{r_p}$$

$$\Leftrightarrow a\widetilde{L}_E = r^2 + \ell^2 - \left. \frac{4r\Delta_r}{\Delta_r'} \right|_{r_p}. \tag{4.14}$$

Because the shadow of a rotating black hole ($a \neq 0$) seen by the standard observer is always symmetric with respect to a horizontal axis, there have to be values $r_{p_\pm}$ such that $\psi(r_{p_\pm}) = \pm\pi/2$. For these parameters $r_{p_\pm}$, (4.12b) yields with (4.14)

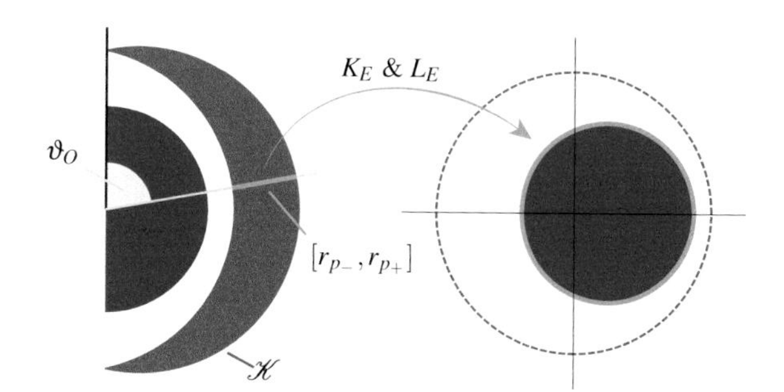

Fig. 4.3 Illustration of the parametrization of the shadow's boundary curve

$$\left(\Sigma\Delta_r' - 4r\Delta_r \mp 4ar\sqrt{\Delta_r\Delta_\vartheta}\,\sin\vartheta\right)\Big|_{(r_{p\pm},\vartheta_O)} = 0. \tag{4.15}$$

Comparison with the inequality (3.7) shows that the radii $r_{p_\pm}$ are in the intersection of the boundary of the exterior photon region $\mathcal{K}$ and the cone $\vartheta = \vartheta_O$. Thus, substituting K_E and $\widetilde{L}_E$ in (4.12a, 4.12b) by the expressions (4.14) provides the shadow's boundary curve $\big(\theta(r_p), \psi(r_p)\big)$ where $r_p \in [r_{p_-}, r_{p_+}]$, see also Fig. 4.3.

The light rays that correspond to the boundary of the shadow are independent of the observer's state of motion. Thus, in the general case of a moving observer, the boundary curve of the shadow is given in the same way as for the standard observer. This means that the shadow's boundary curve $\big(\theta(r_p), \psi(r_p)\big)$—now specified by (4.8a, 4.8b) with K_E and L_E replaced according to (4.14)—is parametrized by $r_p \in I$ where I is the intersection of the exterior photon region $\mathcal{K}$ with the cone $\vartheta = \vartheta_O$ as before.

If $a = 0$, it is not possible to parametrize the boundary curve with r_p because the photon region $\mathcal{K}$ degenerates into a photon sphere $r = r_p$ with unique r_p given by (3.7). By (4.14), the distinct r_p defines a unique $K_E(r_p)$ but does not restrict $\widetilde{L}_E$ (or L_E). Calculating the corresponding constant θ from (4.8a, 4.8b) or (4.12a, 4.12b) gives the radius of the shadow which is circular in this case. Thus, the boundary curve of the shadow has the form $\big(\theta(\widetilde{L}_E), \psi(\widetilde{L}_E)\big)$ where $\widetilde{L}_E$ ranges between the extremal values determined by (3.3c) for $\Theta(\vartheta_O) = 0$

$$L_E = \chi \pm \sqrt{\Delta_\vartheta K_E}\,\sin\vartheta = \chi \pm 4r\sin\vartheta\,\frac{\sqrt{\Delta_\vartheta\Delta_r}}{\Delta_r'}\Bigg|_{(r_p,\vartheta_O)} \tag{4.16a}$$

$$\Longleftrightarrow \quad \widetilde{L}_E = -a\cos^2\vartheta - 2\ell\cos\vartheta \pm 4r\sin\vartheta\,\frac{\sqrt{\Delta_\vartheta\Delta_r}}{\Delta_r'}\Bigg|_{(r_p,\vartheta_O)} \tag{4.16b}$$

Before plotting some shadows of black holes let's summarize some general properties of the shadow.

The photon region as well as the boundary of the shadow are described by identical formulas, (3.7) or (4.8a, 4.8b) and (4.12a, 4.12b), for the whole Plebański–Demiański class of black hole space-times. However,the involved metric functions (2.2) have

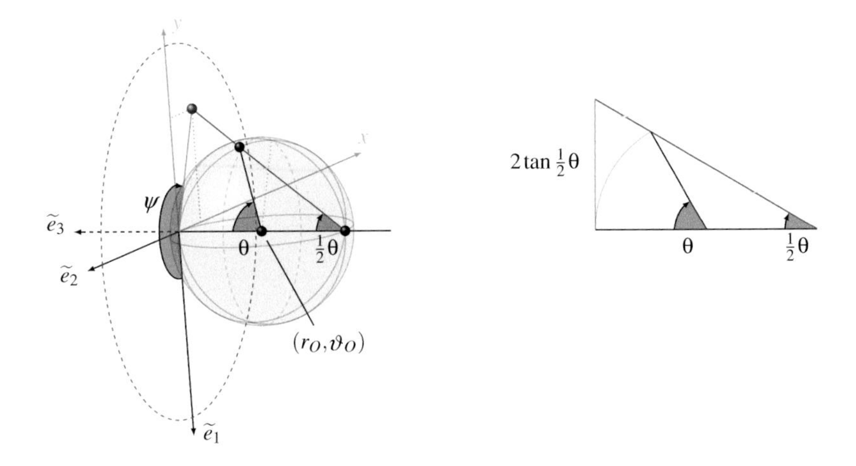

Fig. 4.4 Relation between celestial coordinates (θ, ψ) and standard cartesian coordinates (x, y) is given by stereographic projection, see Eq. (4.17)

different meanings. The shadow seen by the standard observer ($\mathbf{v} = 0$) is always symmetric with respect to a horizontal axis and independent of the Manko–Ruiz parameter C since $\widetilde{L}_E$ determined by Eq. (4.16b) does not depend on C. For an observer in a different state of motion ($\mathbf{v} \neq 0$) this is in general not the case. Here, the observed shadow is distorted by aberration and the distortion is in principle given by the standard aberration formula of special relativity. As the aberration formula maps circles onto circles, the statement that a non-rotating black hole produces a circular shadow is true for an observer in *any* state of motion. Even with non-zero acceleration parameter, which breaks the spherical symmetry, these properties are preserved.

The following sections are filled with several plots of shadows for various black hole space-times. For visualization, the shadows are mapped from the celestial sphere onto a plane by stereographic projection, as illustrated in Fig. 4.2. Standard Cartesian coordinates in that plane are given by, see Fig. 4.4

$$\begin{pmatrix} x(\rho) \\ y(\rho) \end{pmatrix} = -2\tan\left(\tfrac{1}{2}\theta(\rho)\right) \begin{pmatrix} \sin\psi(\rho) \\ \cos\psi(\rho) \end{pmatrix}. \tag{4.17}$$

Consequently, in the projection, i.e., in the xy plane, the θ coordinate can be regarded as radius while the ψ coordinate is a polar angle.

4.1 Kerr–Newman–NUT Space-Times

In the Figs. 3.4, 3.5 and 3.6, we showed plots of the black hole surrounding photon region for the Kerr–Newman, the Kerr–NUT and the Kerr–Newman–NUT class of space-times. The corresponding images of the shadows seen by the standard observer are comprised in Fig. 4.5. Here, the observer position is kept fixed at Boyer–Lindquist coordinates $r_O = 5m$ and $\vartheta_O = \pi/2$ which is always located in the domain of outer communication, cf. Table A.2.

Each of the shadings corresponds to a certain choice of black hole parameters β and ℓ, and for each choice the shadow is shown for four different values of the spin, $a = \lambda a_{\max}$, where $a_{\max} = \sqrt{m^2 + \ell^2 - \beta}$. As for the photon regions, the shadow of a Kerr black hole is plotted for each space-time class for the purpose of comparison where a class is represented by two exemplary sets of parameters; these are the same as for the photon regions in Sect. 3.1.

We see that the shape of the shadow is largely determined by the spin a of the black hole. With increasing a, the shadow becomes more and more asymmetric with respect to a vertical axis and develops almost always a D-shaped contour for a maximally rotating black hole; this is a nice description used by James et al. (2015). The asymmetry is well-known from the Kerr metric and it is easily understood as a "dragging effect" of the rotating black hole on the light rays. Only for highly charged black holes, the shadow stays roughly circular at increasing spin, even in the extreme case as shown in Fig. 4.7. The reason is that $a_{\max} = \sqrt{m^2 - \beta}$ decreases with increasing charge β. Therefore, highly charged and maximally rotating black holes are similar to slowly rotating but hardly charged black holes, cf. Sect. 3.1. To highlight this aspect, Fig. 4.6 contains all possible shadow images for four Kerr–Newman black holes for the given fixed spin values; plots in the "empty triangle" would, of course, belong to naked singularities. For the same reason, maximally spinning black holes in more general space-time classes will lose the D-shaped contour, too, if they are sufficiently charged.

Such an effect is not observable in the Kerr–NUT space-time since the spin of an extreme black hole, $a_{\max} = \sqrt{m^2 + \ell^2}$, increases with increasing gravitomagnetic NUT charge ℓ. This has only small effect on the shadow in contrast to the photon sphere which is much more influenced by ℓ than by β (in view of the symmetry loss); ℓ affects the size of the shadow but hardly its shape, at least for the naked eye. Interestingly, the shadow stays symmetric in spite of the non-symmetric photon region. Since the shadow seen by the standard observer is independent of the Manko–Ruiz parameter C (consider equations with $\widetilde{L}_E$), we refer to Sect. 4.4 for images corresponding to the photon regions in Fig. 3.8; in the mentioned section, we analyze C-influenced shadows seen by moving observers.

Furthermore, we should mention that in the case $\ell \neq 0$ some light rays have to pass through the singularity on the axis. We have assumed that these light rays are *not* blocked, i.e., that the source of the gravitomagnetic NUT field does *not* cast a shadow.

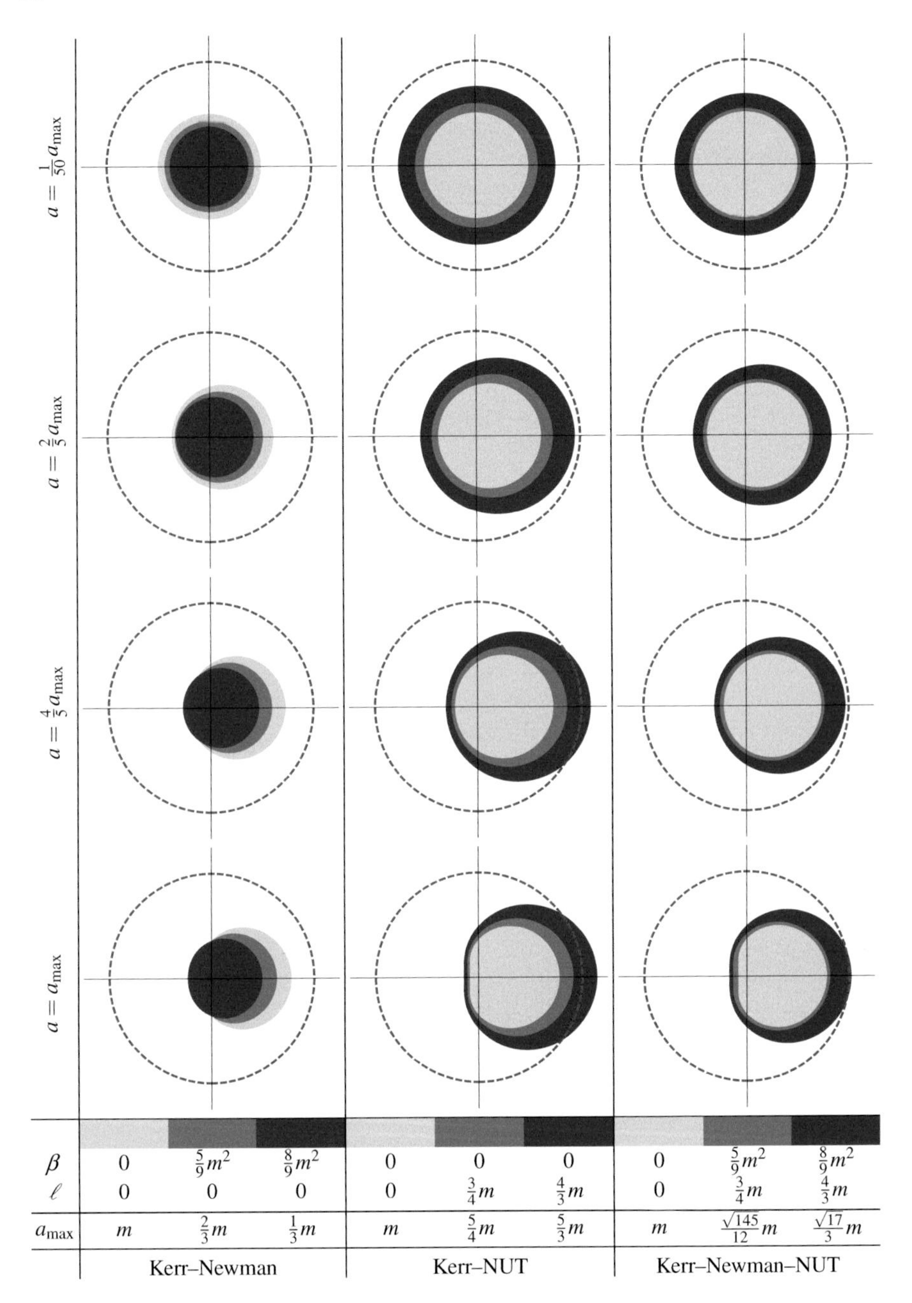

Fig. 4.5 Shadows of black holes in Kerr–Newman–NUT space-times for different parameters a, β, ℓ seen by an observer at $r_O = 5\,\mathrm{m}$ and $\vartheta_O = \pi/2$, cf. Table A.2. The *dashed* (*red*) *circle* indicates the celestial equator, cf. Fig. 4.2

Note that the size of the shadow depends, of course, on r_O and that there is no direct way of comparing radial coordinates in different space-times operationally. Therefore, if we want to get some information on the space-time from observing the shadow, the shape is much more relevant than the size. In summary, the standard observer sees a symmetric shadow of a Kerr–Newman–NUT black hole whose size depends on β and ℓ but not on the spin in contrast to the shape.

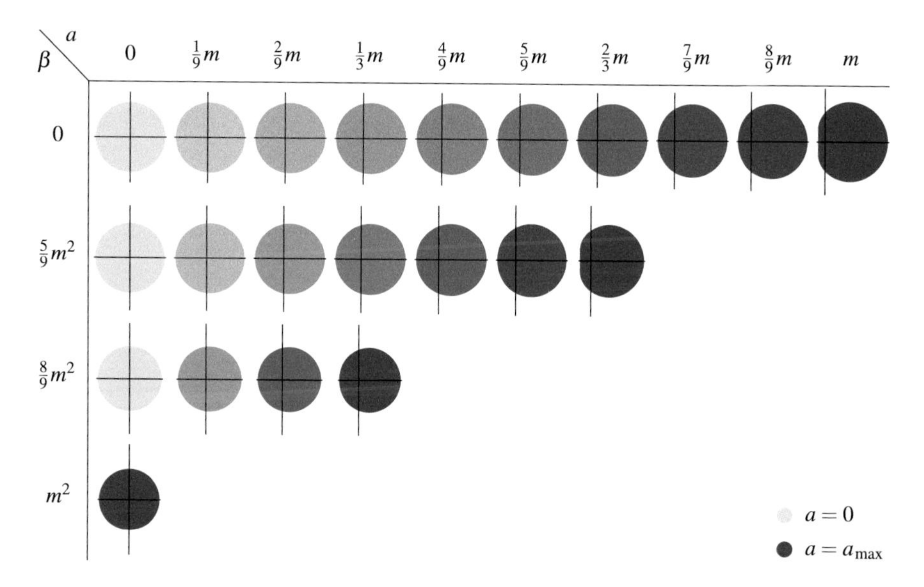

Fig. 4.6 Shadows of Kerr–Newman black holes with fixed spin parameters a, cf. Table A.4. The color density of the plots corresponds to the ratio $\frac{a}{a_{\max}} = \frac{a}{\sqrt{m^2-\beta}}$

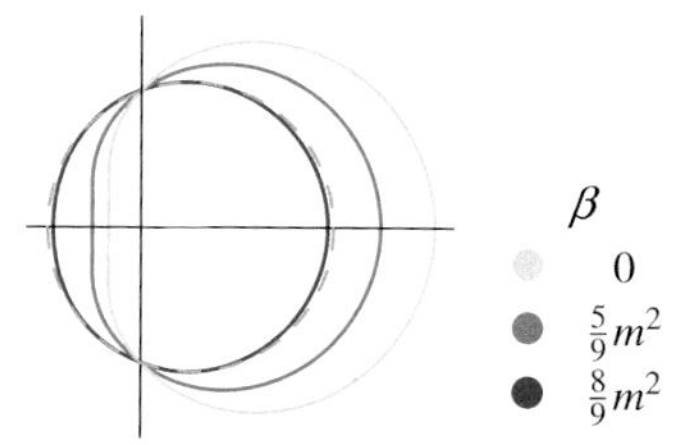

Fig. 4.7 Contour of the black hole's shadow in extremal Kerr–Newman space-times presented in Fig. 4.5. The overlaid *green dashed circle* shows that the shadow of the highly charged black hole ● is almost circular

4.2 Space-Times with Cosmological Constant

In Fig. 4.8, the shadows of black holes in the Plebański class of space-times are presented, i.e., shadows of Kerr–Newman–NUT black holes with cosmological constant. Again, the standard observer is located at fixed Boyer–Lindquist coordinates $r_O = 5m$ and $\vartheta_O = \pi/2$ in the domain of outer communication, cf. Table A.5. Since the spin affects the shadow in the same way as for $\lambda = 0$, we consider only maximally rotating black holes. The images show that the effects of the cosmological constant Λ on the shadow are marginal as it is the case for the photon regions. Only relatively large values of Λ change the shadow noticeable but just in size. A positive cosmological constant acts size reducing but changes in the shape are not recognizable. The most remarkable effect is that the shadows are still symmetric with respect to a horizontal axis.

4.3 Accelerated Black Holes

The shadows of maximally rotating black holes in accelerated space-times are shown in Fig. 4.9. From Table A.6 we can read off that for the chosen parameters the smallest value for the cosmological horizon is $r_4 = 2.69\,\text{m}$ in the Kerr–NUT space-time. Therefore, the standard observer is located at fixed Boyer–Lindquist coordinates $r_O = 2.6\,\text{m}$ and $\vartheta_O = \pi/2$ in the domain of outer communication. The different values of α and Λ are encoded into different shadings.

Also with acceleration, the shape of the shadow is largely determined by the spin a of the black hole. Hence, the shadow becomes more and more asymmetric with respect to a vertical axis with increasing spin a where the asymmetry results from the "dragging effect" of the rotation on the light rays. In view of the small changes at the special regions this effect is expectable. But there are also non-trivial results: Although the acceleration parameter breaks the spherical symmetry, a non-rotating black hole still has a circular shadow since (4.8a) depends on the unique $K_E(r_p)$ but not on $\widetilde{L}_E$; consequently, $\theta =$ constant which is why the shadow is circular. Independent of the rotation, the shadow is—as in the non-accelerated case—always symmetric with respect to a horizontal axis since (ψ, θ) and $(\pi - \psi, \theta)$ are determined by the same constants of motion K_E and $\widetilde{L}_E$. Because this result is not implied by an underlying symmetry unless $\ell = 0$, $\alpha = 0$ and $\vartheta_O = \pi/2$, it is non-trivial. Furthermore, the shadow is still independent of the Manko–Ruiz parameter C which is relevant only in the case $\ell \neq 0$. Thus, several properties of the shadow are preserved, even with added acceleration parameter. The acceleration has an effect on the *size* of the shadow, as is visible with the naked eye. This, however, has little relevance in view of observations because the size also scales with r_O and a comparison of the radial coordinates in different space-times has no direct operational meaning.

Shadow images for different signs of a, ℓ and α are shown in Fig. 4.10. Here, the standard observer is in almost all cases located at $r_O = 2.6\,\text{m}$ and $\vartheta_O = \pi/2$ as

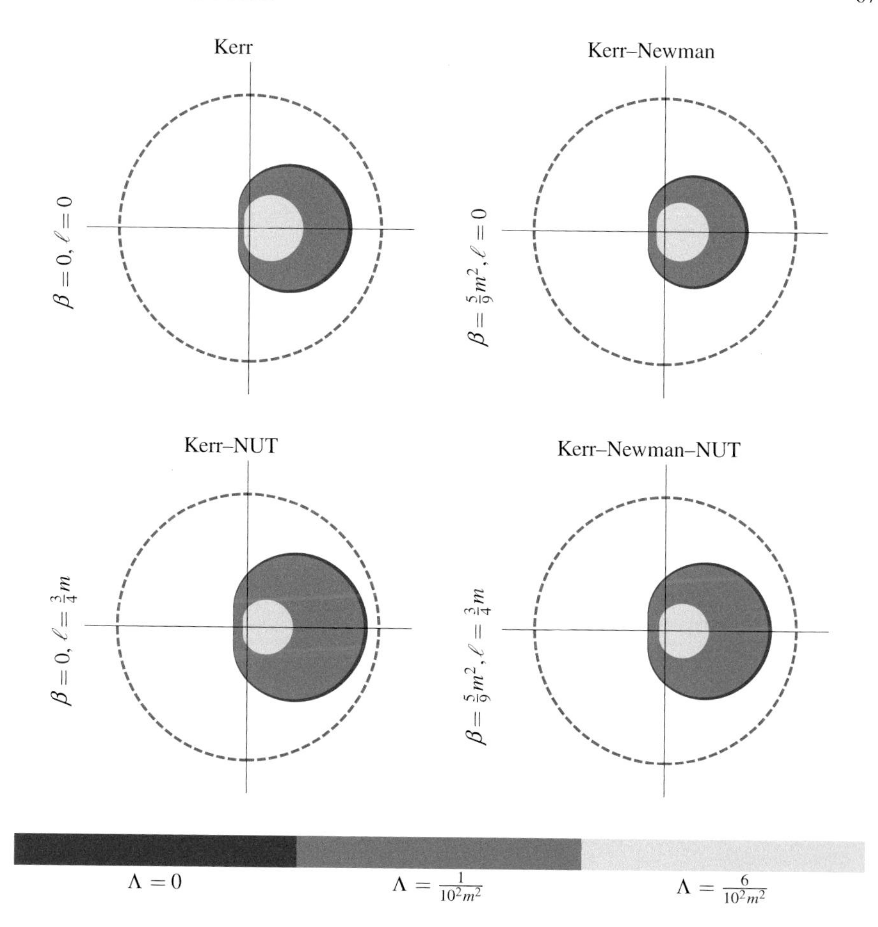

Fig. 4.8 Influence of the cosmological constant Λ on the shadow of extremal Kerr–Newman–NUT black holes seen by an observer at $r_O = 5\,\text{m}$, $\vartheta_O = \pi/2$, cf. Table A.5. The plots of shadows have different parameters β, ℓ and Λ $(C = 0)$ with color-coded magnitude of Λ. The *dashed* (*red*) *circle* indicates the celestial equator, cf. Fig. 4.2

before. Only if $\alpha\ell < 0$ (last row), the observer has to be fixed at a bigger r coordinate $r_O = 7\,\text{m}$ due to a bigger event horizon, cf. Table A.7. The shadow is reflected at a vertical axis if the sign of a, i.e., the spin direction, is changed. One might have expected a similar effect with respect to a horizontal axis if the sign of α is changed. However, this is not true. As the shadow stays symmetric with respect to a horizontal axis even if $\alpha \neq 0$, the shadow is independent of the direction of the acceleration, i.e., of the sign of α.

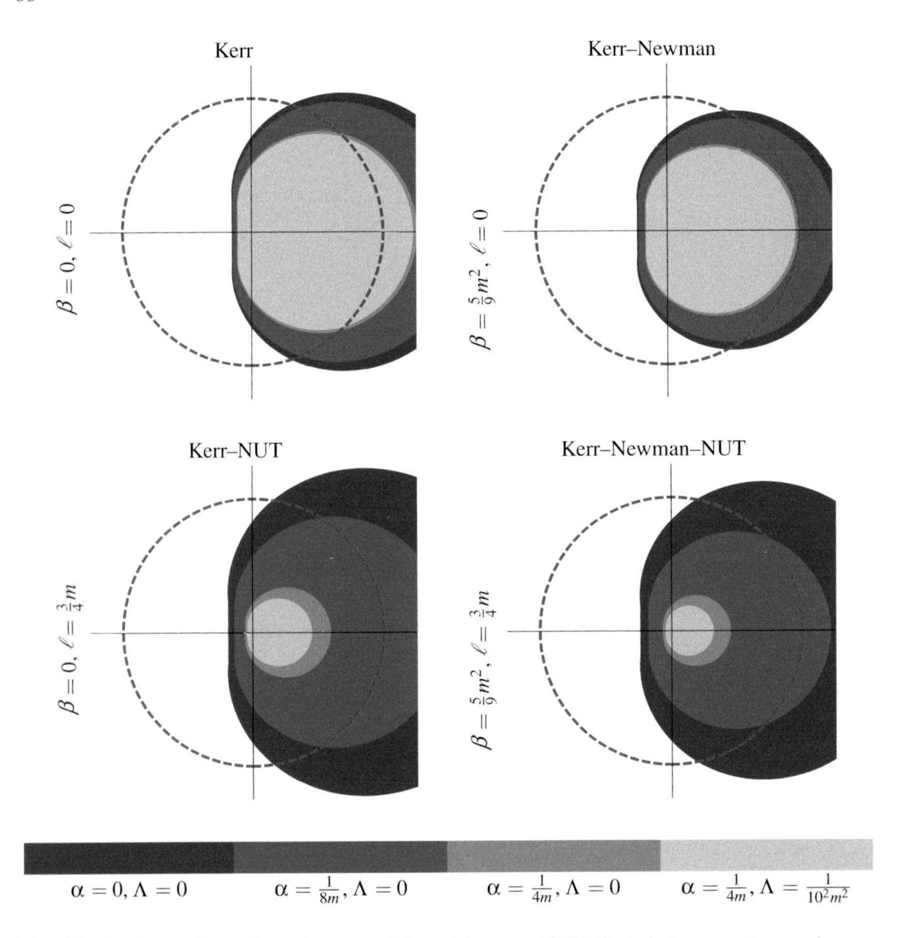

Fig. 4.9 Shadows of accelerated extremal Kerr–Newman–NUT black holes seen by an observer at $r_O = 2.6\,\mathrm{m}$ and $\vartheta_O = \pi/2$, cf. Table A.6. The plots of shadows have different parameters β, ℓ, Λ and α ($C = 0$) with color-coded magnitude of α. The *dashed* (*red*) *circle* indicates the celestial equator, cf. Fig. 4.2

4.4 Moving Observers

Up to now, we discussed the influence of the different black hole parameters on the shadow of black holes seen by the standard observer. In the following, we examine what an observer moving relatively to the standard observer would see. As described before, we use our analytical parameter representation (4.8a, 4.8b) with (4.10) and (4.14) to calculate the boundary curve of the shadow as seen by an observer moving with four-velocity $\widetilde{e}_0$. The results in Fig. 4.11 are visualized via stereographic projection from the celestial sphere onto a plane; see Figs. 4.2 and 4.4 for illustrations and Eq. (4.17) for standard Cartesian coordinates in this plane. Each subfigure combines the pictures for four spin values $a = \lambda a_{\max}$.

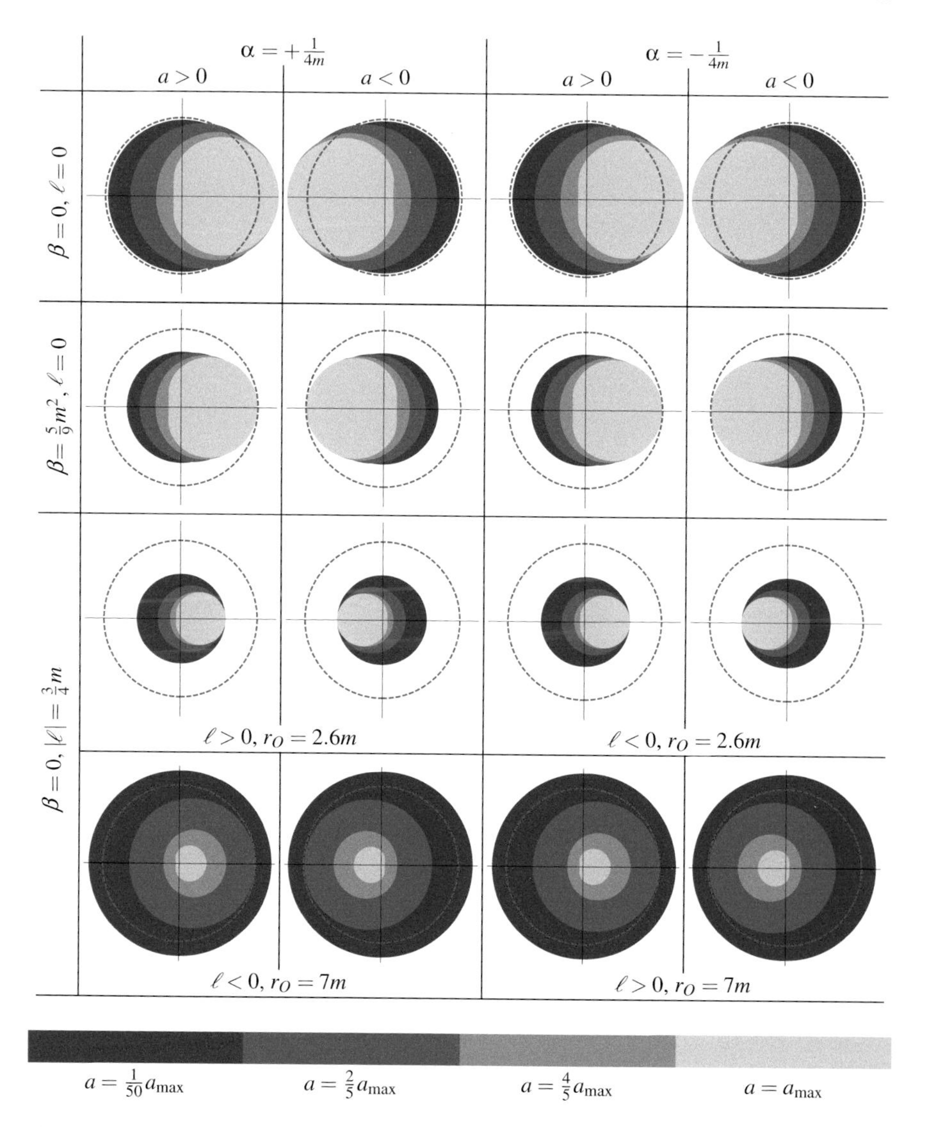

Fig. 4.10 Shadows of black holes for different signs of the spin a, the NUT charge ℓ ($C = 0$) and the acceleration α seen by an observer at $r_O = 2.6\,\mathrm{m}$ or $r_O = 7\,\mathrm{m}$ (only *last row*) and $\vartheta_O = \pi/2$. The magnitude of a is color-coded and values for $|a_{\max}|$ are noted in Table A.7

The green star represents the direction of the observer's motion: It is drawn as stereographic projection, Eq. (4.17), of the point $\{\theta_v, \psi_v\}$ that is given by the usual transformation from Cartesian to spherical coordinates

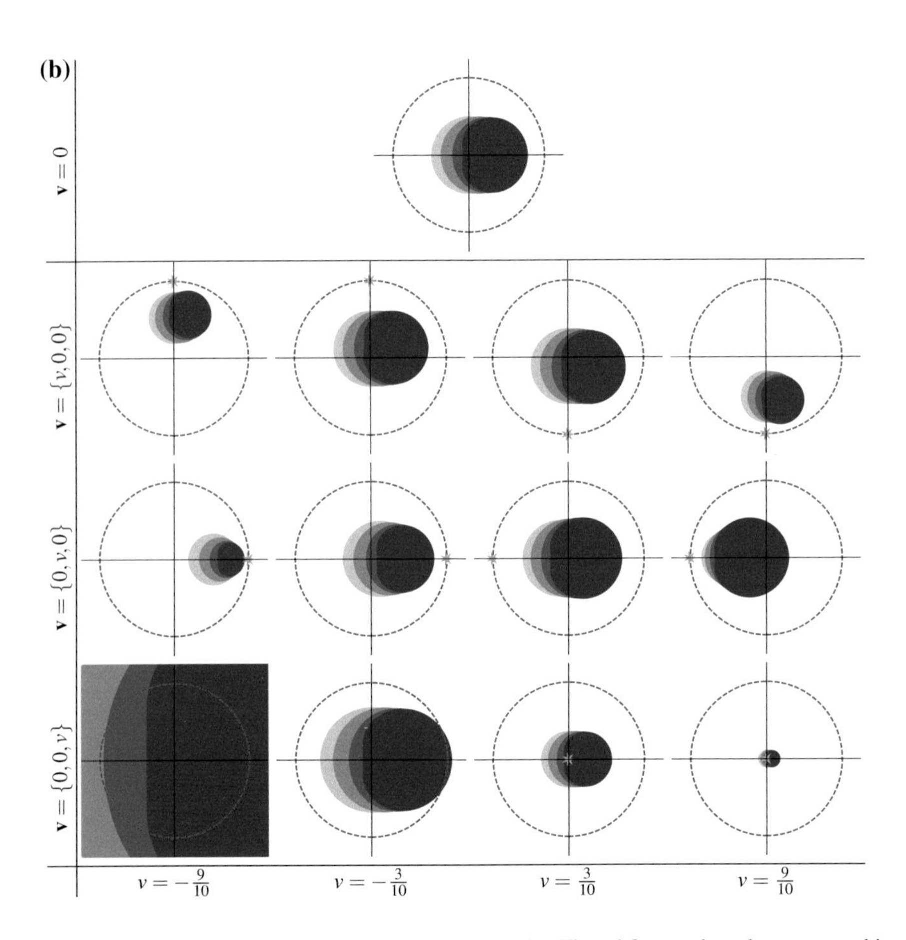

Fig. 4.11 Aberrational effects on the shadows of black holes. The subfigures show the stereographic projection of the shadow for observers ($r_O = 5\,\mathrm{m}$, $\vartheta_O = \frac{\pi}{2}$) moving with various velocities $\mathbf{v} = (v_1, v_2, v_3)$. The projected direction of the observer's motion is marked by a (*green*) star ✳. Each plot combines the silhouettes for four different spins $a = \lambda a_{\max}$, see Table A.6 for the corresponding values of $a_{\max}$. **a** Color code for spin values, **b** fixed Kerr space-time, varying $\mathbf{v} = (v_1, v_2, v_3)$, **c** various space-times, fixed $\mathbf{v} = \left(\frac{3}{10}, -\frac{3}{10}, -\frac{1}{10}\right)$

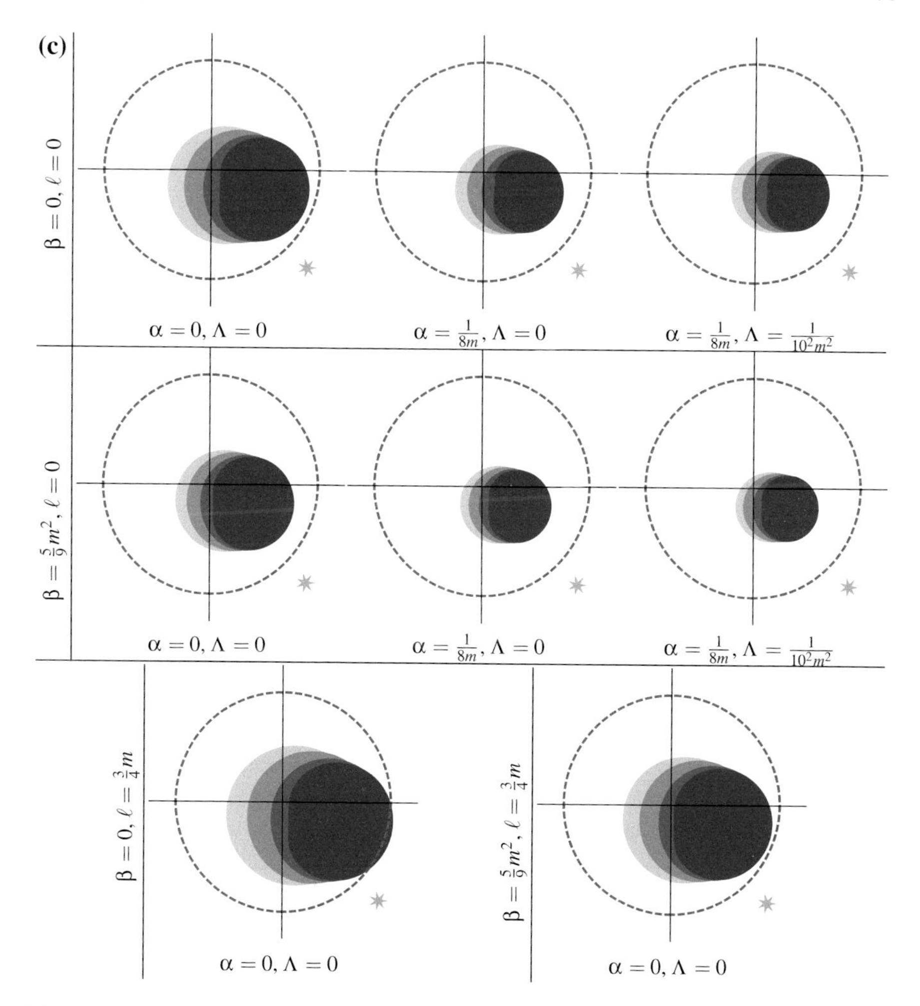

Fig. 4.11 (continued)

$$\tan\theta_v = \frac{\sqrt{v_1^2 + v_2^2}}{v_3}, \qquad \tan\psi_v = \frac{v_2}{v_1}. \tag{4.18}$$

In case of a pure radial motion with $\mathbf{v} = \{0, 0, v_3\}$, we place the star in the origin if the observer moves towards the black hole ($v_3 > 0$) and omit plotting the star otherwise.

In principle, the shadows for moving observers ($\mathbf{v} \neq 0$) are calculable from the shadow seen by the standard observer ($\mathbf{v} = 0$) with the help of Penrose's aberration formula, cf. Penrose (1959)

$$\tan\frac{\widetilde{\alpha}}{2} = \sqrt{\frac{c-v}{c+v}}\tan\frac{\alpha}{2}. \tag{4.19}$$

But for applying this formula, one may need coordinate transformations since the angles α and $\widetilde{\alpha}$ have to be measured against the direction of the motion. Hence, no transformations are needed if the observer moves in radial direction. Then, the shadow is magnified if the observer moves away from the black hole, and demagnified if the observer moves towards the black hole. In this case, our formula (4.8a) reduces to the following common variant of Penrose's aberration formula (4.19)

$$\cos\widetilde{\theta} = \frac{v + \cos\theta}{1 + v\cos\theta}. \tag{4.20}$$

Penrose (1959) emphasized in his article that the aberration formula maps circles on the celestial sphere onto circles. Thus, the shadow of a non-rotating black hole ($a = 0$) is always circular, independent of the observer's motion. Consequently, our pictures of the shadow are always circular then, because the stereographic projection (4.17) maps circles onto circles, too.

Figure 4.11b shows several pictures of shadows for differently moving observers in Kerr space-times. The result for the standard observer ($\mathbf{v} = 0, \widetilde{e}_\mu = e_\mu$) is shown in the top row. In each of the lower rows we vary only one component v_i of $\mathbf{v}$; in the following we write $\mathbf{v_i}$ as abbreviation for those observer velocities $\mathbf{v}$ with $v_i \neq 0$ and $v_j = 0, j \neq i$. Due to our definition of the tetrad e_μ in (4.1) and of the observer's four-velocity $\widetilde{e}_0$ in (4.2a), the observer moves in ϑ direction if $\mathbf{v} = \mathbf{v_1}$, and in r, i.e., radial direction if $\mathbf{v} = \mathbf{v_3}$. For $\mathbf{v} = \mathbf{v_2}$ the motion is in φ direction.

In the last row of Fig. 4.11b, the shadow images seen by a radially moving observer ($\mathbf{v} = \mathbf{v_3}$) are shown; it can be seen that the shadows are magnified if the observer moves away from the black hole (v_3 negative), and demagnified, if the observer moves towards the black hole (v_3 positive), as mentioned before.

For velocities $\mathbf{v} = \mathbf{v_1}$ or $\mathbf{v} = \mathbf{v_2}$, the shadow is shifted in the direction of the observer's motion while the effect increases with the velocities. Also the size of the shadow is affected. But all these aberrational changes are explainable if one relates the direction of the observer's motion to the spin of the black hole and to the equatorial plane as symmetry plane.

Furthermore, the shadow is symmetric with respect to a horizontal axis as long as the observer does not move in ϑ direction because, as in (4.8b), $\sin\psi$ depends on $\dot{\vartheta}$ which is given by a quadratic expression, see (3.3c). Hence, the different signs of $\dot{\vartheta}$ yield different solutions of (4.8a, 4.8b) for the points (θ, ψ) and $(\theta, \pi - \psi)$. Without a ϑ component in the velocity, the symmetry of the shadow is not affected even if the observer is not in the equatorial plane, i.e. $\vartheta_O \neq \frac{\pi}{2}$.

The remaining plots in Fig. 4.11c are shadow images calculated for different space-times but with the same observer at $(r_O, \vartheta_O) = (5m, \frac{\pi}{2})$ moving with velocity $\mathbf{v} = \left(\frac{3}{10}, -\frac{3}{10}, -\frac{1}{10}\right)$. We see that all images are affected in the same way by the velocity $\mathbf{v}$, i.e., all shadows are shifted in the direction of the $\mathbf{v}$. But besides this shift, no further effects coming from the space-time are visible.

All in all, the shadows shown in Fig. 4.11 are calculated for relatively fast moving observers ($v = 0.3\,c$ up to $v = 0.9\,c$). Thus, the aberrational influence for the future observations of the shadow of Sgr A* within the Event Horizon Telescope or the BlackHoleCam project is expected to be very small since our solar system orbits the Galactic center with roughly $250\frac{\text{km}}{\text{s}} \approx \frac{1}{1000}c$, see Reid et al. (2009). Nevertheless, the study of aberrational effects are of interest from a fundamental point of view.

Zero Angular Momentum Observer (ZAMO). We have seen that the Manko–Ruiz parameter C has no influence on the shadow seen by the standard observer independent of the influence on the photon region, see Fig. 3.8. This is in general not true for a moving observer. If the observer moves with velocity $\mathbf{v} = \Big(0, \frac{\sqrt{\Delta_r}\chi}{(\Sigma+a\chi)\sqrt{\Delta_\vartheta}\sin\vartheta}, 0\Big)$ we find for (4.2a, 4.2b)

$$\widetilde{e}_0 = \frac{g_{\varphi\varphi}\partial_t - g_{\varphi t}\partial_\varphi}{\sqrt{g_{\varphi\varphi}(g_{\varphi t}^2 - g_{\varphi\varphi}g_{tt})}}, \quad \widetilde{e}_1 = \frac{1}{\sqrt{g_{\vartheta\vartheta}}}\,\partial_\vartheta, \quad \widetilde{e}_2 = \frac{-1}{\sqrt{g_{\varphi\varphi}}}\,\partial_\varphi, \quad \widetilde{e}_3 = \frac{-1}{\sqrt{g_{rr}}}\,\partial_r. \tag{4.21}$$

Here, the spatial vectors $\widetilde{e}_i$ are just the normalized standard basis vectors ∂_i and the observer falls with four velocity $\widetilde{e}_0$ towards the black hole. By definition, we have ${\widetilde{e}_0}^{\varphi} = \dot{\varphi}$ and ${\widetilde{e}_0}^{t} = \dot{t}$ which is why $(\widetilde{e}_0)_\varphi$ is the observer's angular momentum L_O, cf. Eq. (3.2). But since L_O vanishes,

$$L_O = (\widetilde{e}_0)_\varphi = g_{\varphi\varphi}{\widetilde{e}_0}^{\varphi} + g_{\varphi t}{\widetilde{e}_0}^{t} \overset{4.21}{=} \frac{-g_{\varphi\varphi}g_{\varphi t} + g_{\varphi t}g_{\varphi\varphi}}{\sqrt{g_{\varphi\varphi}(g_{\varphi t}^2 - g_{\varphi\varphi}g_{tt})}} = 0, \tag{4.22}$$

this observer is called *zero angular momentum observer (ZAMO)*.

Comparing Eqs. (4.21) and (4.5) we find $k_{0r} = k_{0\vartheta} = k_{3\vartheta} = 0$, and (4.7a) and (4.7b) reduce to

$$\dot{r} = \sigma k_{3r}\cos\theta, \quad \dot{\vartheta} = \sigma k_{1\vartheta}\sin\theta\cos\psi. \tag{4.23}$$

Apparently, these equations can be solved easily for $\sin\theta$ and $\sin\psi$, much easier than by inserting the observer's velocity $\mathbf{v}$ in (4.8a, 4.8b). With

$$\sigma = \left.\frac{-L_z g_{\varphi t} - E g_{\varphi\varphi}}{\sqrt{g_{\varphi\varphi}(g_{\varphi t}^2 - g_{\varphi\varphi}g_{tt})}}\right|_{(r_O,\vartheta_O)}, \tag{4.24}$$

these calculations yield the following formulas for the shadow's boundary seen by the ZAMO

$$\sin^2\theta = \frac{g_{\varphi t}^2 - g_{\varphi\varphi}g_{tt}}{(g_{\varphi t}L_E + g_{\varphi\varphi})^2}\left(L_E^2 + \frac{\Omega^2}{\Sigma}g_{\varphi\varphi}\left(K_E - \frac{(\chi - L_E)^2}{\Delta_\vartheta \sin^2\vartheta}\right)\right)\Bigg|_{(r_O,\vartheta_O)}, \quad (4.25a)$$

$$\sin^2\psi = \frac{\Sigma L_E^2}{L_E^2 + \frac{\Omega^2}{\Sigma}g_{\varphi\varphi}\left(K_E - \frac{(\chi - L_E)^2}{\Delta_\vartheta \sin^2\vartheta}\right)}\Bigg|_{(r_O,\vartheta_O)}, \quad (4.25b)$$

where L_E and K_E are given by Eq. (4.14) again.

Images of shadows of a Kerr–Newman–NUT black hole seen by the ZAMO are shown in Fig. 4.12. They correspond to the photon regions in Fig. 3.8. Since the ZAMO moves relative to the standard observer in φ direction, the changes of the shadow match those for the Kerr space-time in Fig. 4.11b.

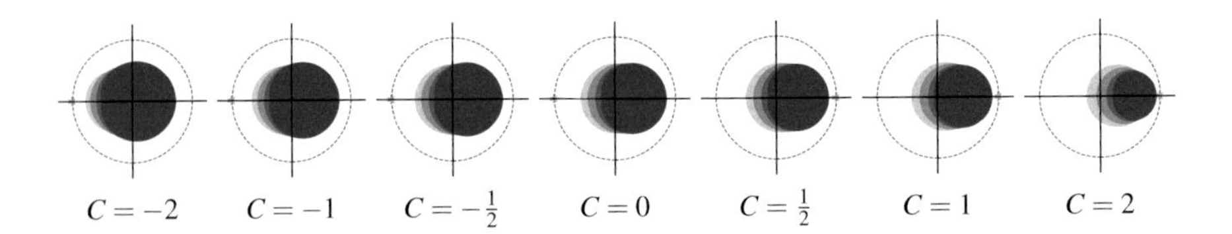

Fig. 4.12 Shadow of a black hole ($\beta = \frac{5}{9}m^2$, $\ell = \frac{4}{3}m$) for varying singularity parameter C seen by the zero angular momentum observer ($r_O = 6m$, $\vartheta_O = \frac{\pi}{2}$); the ZAMO moves with $\mathbf{v} = \left(0, \frac{\sqrt{\Delta_r}\chi}{(\Sigma + a\chi)\sqrt{\Delta_\vartheta}\sin\vartheta}, 0\right)$ relative to the standard observer. As always, the *cross hairs* indicate the spatial direction towards the black hole and the *dashed* (*red*) *circle* indicates the celestial equator

4.5 Inclination of Observer

In Fig. 4.13, we consider a specific Kerr–Newman–NUT black hole with fixed parameters β and ℓ. We keep the radius coordinate r_O of the observer fixed and vary the inclination ϑ_O. Clearly, the dragging asymmetry with respect to the vertical axis vanishes when the observer approaches the axis; in the limit $\vartheta_O \to 0$, the shadow becomes circular because with respect to the rotation axis the black hole is symmetric again. This effect is independent of the observer's state of motion. We already have emphasized the remarkable fact that the shadow is always symmetric with respect to the horizontal axis as long as the observer's motion does not have a ϑ part, i.e., $\mathbf{v} = \{0, *, *\}$.

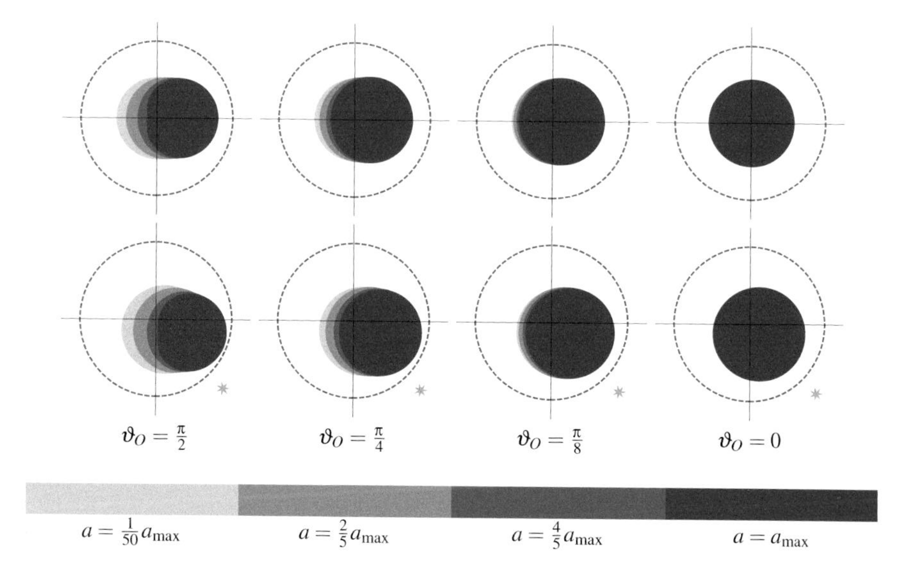

Fig. 4.13 Shadow of a Kerr–Newman–NUT black hole ($\beta = \frac{5}{9}m^2, \ell = \frac{3}{4}m, a_{\max} = \frac{\sqrt{145}}{12}m$) for an observer at $r_O = 5\,\mathrm{m}$ and different inclination angles ϑ_O. The *first row* is calculated for the standard observer and the *second row* for an observer moving with $\mathbf{v} = (\frac{3}{10}, -\frac{3}{10}, -\frac{1}{10})$

4.6 Angular Diameter of the Shadow

From the analytical formulas (4.8a, 4.8b)/(4.12a, 4.12b) and (4.14) for the boundary curve of the shadow, we can deduce expressions for the horizontal and vertical angular diameters of the shadow. They correspond to the dashed lines in Fig. 4.14. Owing to the symmetry, the angular diameters δ_h and δ_v are determined by three angular radii ρ_{h_1}, ρ_{h_2} and ρ_v as indicated in Fig. 4.14,

$$\delta_h = \rho_{h_1} + \rho_{h_2}, \qquad \sin\rho_{h_i} = \sin\psi_{h_i}\sin\theta_{h_i} = P(r_{h_i})T(r_{h_i}), \tag{4.26}$$

$$\delta_v = 2\rho_v, \qquad \sin\rho_v = \cos\psi_v\sin\theta_v = \sqrt{1 - P^2(r_v)}T(r_v), \tag{4.27}$$

where T and P have the same meaning as in (4.12a, 4.12b).

In the following, we restrict ourselves to the Kerr space-time with an observer in the equatorial plane, $\vartheta_O = \frac{\pi}{2}$. Even in this case, a formula for the angular diameters of the shadow was not derived before, as far as we know. In the general case, the angular diameters can be calculated analogously; then it is true that the radius values r_{h_i} and r_v are zeros of a polynomial of higher than fourth order, so they cannot be determined in closed form. In terms of these radii, however, one gets analytical formulas for the angular diameters even in the general case. The horizontal angular radii ρ_{h_i} are characterized by $\psi_{h_i} = \pm\frac{\pi}{2}$, so we must solve the equation $1 = \sin^2\psi(r_h) = P^2(r_h)$ which in the Kerr case simplifies to (use Eq. (4.8b) with Eq. (4.14))

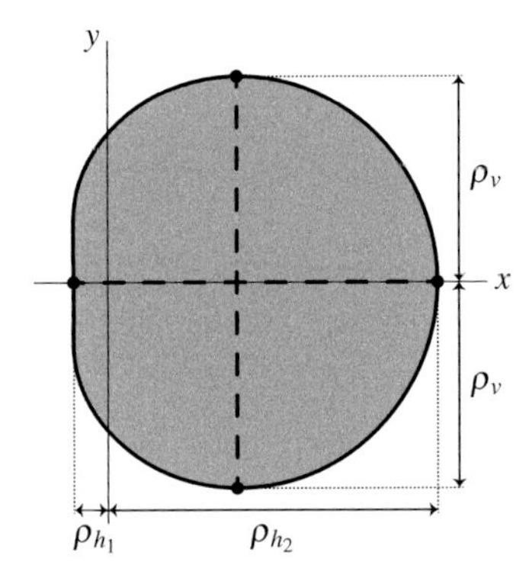

Fig. 4.14 (From [3]) Angular radii of the shadow of a black hole. Owing to the symmetry with respect to a *horizontal axis*, the two angular diameters (*dashed lines*) of the shadow are given by three angular radii: two horizontal radii ρ_{h_i} and one vertical radius ρ_v. The angular diameters are calculated as $\delta_h = \rho_{h_1} + \rho_{h_2}$ and $\delta_v = 2\rho_v$, respectively

$$r_h(r_h - 3m)^2 = 4ma^2 \tag{4.28}$$

$$\Rightarrow \quad r_{h_1} = 2m + 2m\cos(\zeta/3), \tag{4.29a}$$

$$r_{h_2} = 2m - m\cos(\zeta/3) - \sqrt{3}m\sin(\zeta/3), \tag{4.29b}$$

$$r_{h_3} = 2m - m\cos(\zeta/3) + \sqrt{3}m\sin(\zeta/3), \tag{4.29c}$$

where $\zeta = \arg\big((2a^2m - m^3) - i(2am\sqrt{m^2 - a^2})\big)$. Here we have to choose the solutions r_{h_1} and r_{h_2} which are the radii of the two circular photon orbits in the exterior photon region. Evaluating PT for r_{h_1} and r_{h_2} yields by (4.26) the horizontal angular diameter δ_h of the shadow.

The vertical angular radius corresponds to boundary points where the tangent is horizontal. By (4.27) we have $f(r_v) := \sin^2\rho_v = \big(1 - P^2(r_v)\big)T^2(r_v)$, so the tangent is horizontal if $\frac{\mathrm{d}f}{\mathrm{d}r_v}(r_v) = 0$. This yields

$$0 = (1 - P^2)T' - PP'T\,|_{r_v} \tag{4.30}$$

$$= \frac{\sqrt{\Delta(r_O)}\, r_v\left(r_v(2a^2 + r_O^2) - 3mr_O^2\right)\left(a^2m - r_v(3m^2 - 3mr_v + r_v^2)\right)}{a^2\sqrt{\Delta(r_v)}\left(r_v(2a^2 + r_O^2 + r_v^2) - m(r_O^2 + 3r_v^2)\right)^2} \tag{4.31}$$

where we have to choose the unique solution inside the exterior photon region

$$r_v = \frac{3mr_O^2}{2a^2 + r_O^2}. \tag{4.32}$$

With this value r_v, we get an analytic expression of the vertical angular radius

$$\sin^2 \rho_v = (1 - P^2)T^2|_{r_v} = \frac{27m^2 r_O^2 \big(a^2 + r_O(r_O - 2m)\big)}{r_O^6 + 6a^2 r_O^4 + 3a^2(4a^2 - 9m^2)r_O^2 + 8a^6}. \tag{4.33}$$

For $a = 0$, we recover from (4.33) Synge's formula (Synge 1966) for a Schwarzschild black hole,

$$\sin^2 \rho = \frac{27m^2(r_O - 2m)}{r_O^3}. \tag{4.34}$$

It was already noted as Eq. (1.2) in the introduction. Since the shadow of a non-rotating black hole is always circular, the horizontal angular radii ρ_{h_i} are also given by Eq. (4.34) in this case.[4] Note that for all values $0 \le a^2 \le m^2$ Eq. (4.33) gives the same result as (4.34), $27m^2/r_O^2$, if m is negligibly small in comparison to r_O. This means that for observers far away from the black hole the vertical diameter of the shadow is independent of a.

In the extremal Kerr space-time, $a = m$, the circular photon orbits are at $r_{h_1} = 4m$ and $r_{h_2} = m$ since $\zeta = \arg(m^3) = 0$. Together with (4.32), this results in the following formulas for the angular radii

$$\sin^2 \rho_{h_1} = \frac{64m^2(r_O - m)^2}{(r_O^2 + 8m^2)^2}, \quad \sin^2 \rho_{h_2} = \frac{m^2}{(r_O + m)^2}, \qquad \sin^2 \rho_v = \frac{27m^2 r_O^2}{(r_O + m)^2(r_O^2 + 8m^2)}. \tag{4.35}$$

Finally, we use (4.34) and (4.35) to determine the angular diameters given by (4.26) and (4.27) for the shadow of the black hole in the center of our Galaxy near Sgr A* and of that in M87. The resulting values are given in Table 4.1 together with the corresponding values for the mass M (in multiples of the Solar mass $M_\odot$) and for the distance r_O of the black holes. We use two sets of parameters for M87 because the mass estimation based on the modeling of stellar dynamics yields a mass twice as big as the estimation based on gas dynamical measurements, compare Broderick et al. (2015), Kormendy and Ho (2013), Gebhardt et al. (2011), Walsh et al. (2013), see Gillessen et al. (2009) for mass and distance of Sgr A*.

The horizontal angular diameter for the maximally rotating black holes is always about 13 % smaller than for the Schwarzschild case while the vertical angular diameters δ_v coincide in all cases. We already observed that the latter is a consequence of the fact that r_O is large in comparison to m. It turns out that the shadow of the black hole in M87 is not much smaller than that of the black hole at the center of our Galaxy; the bigger distance of M87 is almost compensated by its bigger mass.

[4] For $a = 0$, one finds $\zeta = \arg(-m^3) = -\pi$ and $r_{h_{1,2}} = 3\,\mathrm{m}$. Then $T^2(3m)$ reproduces (4.34).

Table 4.1 Horizontal and vertical angular diameter δ_h, δ_v of the shadow for Sgr A* and M87 for a non-rotating Schwarzschild model ($a = 0$) or an extreme Kerr model ($a = m$) of their black holes

	Sgr A*		M87		M87	
	δ_h	δ_v	δ_h	δ_v	δ_h	δ_v
$a = 0$	53.1 μas	53.1 μas	37.8 μas	37.8 μas	20.1 μas	20.1 μas
$a = m$	46.0 μas	53.1 μas	32.8 μas	37.8 μas	17.4 μas	20.1 μas
$m = \frac{MG}{c^2}$	$M = 4.31 \times 10^6 M_\odot$, $r_O = 8.33\,\text{kpc}$		$M = 6.2 \times 10^9 M_\odot$, $r_O = 16.68\,\text{Mpc}$		$M = 3.5 \times 10^9 M_\odot$, $r_O = 17.9\,\text{Mpc}$	

By [1–3] I refer to my papers Grenzebach et al. (2014), Grenzebach (2015) and Grenzebach et al. (2015), respectively. Sentences marked with [i] can be found in total or only slightly modified in the ith paper

References

Broderick AE, Narayan R, Kormendy J, Perlman ES, Rieke MJ, Doeleman SS (2015) The event horizon of M87. Astrophys J 805(2): doi:10.1088/0004-637X/805/2/179. arXiv:1503.03873

Gebhardt K, Adams J, Richstone D, Lauer TR, Faber SM, Gültekin K, Murphy J, Tremaine S (2011) The black hole mass in M87 from GEMINI/NIFS adaptive optics observations. Astrophys J 729:119(13). doi:10.1088/0004-637X/729/2/119

Ghez AM, Salim S, Weinberg NN, Lu JR, Do T, Dunn JK, Matthews K, Morris MR, Yelda S, Becklin EE, Kremenek T, Milosavljevic M, Naiman J (2008) Measuring Distance and Properties of the Milky Way's Central Supermassive Black Hole with Stellar Orbits. Astrophys J 689(2):1044–1062. doi:10.1086/592738

Gillessen S, Eisenhauer F, Trippe S, Alexander T, Genzel R, Martins F, Ott T (2009) Monitoring Stellar Orbits around the Massive Black Hole in the Galactic Center. Astrophys J 692(2):1075–1109. doi:10.1088/0004-637X/692/2/1075

Grenzebach A, Perlick V, Lämmerzahl C (2014) Photon regions and shadows of Kerr–Newman–NUT Black Holes with a cosmological constant. Phys Rev D 89:124,004(12). doi:10.1103/PhysRevD.89.124004. arXiv:1403.5234

Grenzebach A (2015) Aberrational effects for shadows of black holes. In: Puetzfeld et al Proceedings of the 524th WE-Heraeus-Seminar "Equations of Motion in Relativistic Gravity", held in Bad Honnef, Germany, 17–23 Feb 2013, pp 823–832. doi:10.1007/978-3-319-18335-0_25, arXiv:1502.02861

Grenzebach A, Perlick V, Lämmerzahl C (2015) Photon regions and shadows of accelerated black holes. Int J Mod Phys D 24(9):1542,024(22). doi:10.1142/S0218271815420249 ("Special Issue Papers" of the "7th Black Holes Workshop", Aveiro, Portugal, arXiv:1503.03036)

Griffiths JB, Podolský J (2009) Exact space-times in Einstein's general relativity. Cambridge Monographs on Mathematical Physics. Cambridge University Press, Cambridge. doi:10.1017/CBO9780511635397

James O, von Tunzelmann E, Paul F, Thorne KS (2015) Gravitational lensing by spinning black holes in astrophysics, and in the movie Interstellar. Class Quantum Grav 32(6):065,001(41). doi:10.1088/0264-9381/32/6/065001

Kormendy J, Ho LC (2013) Coevolution (or not) of supermassive black holes and host galaxies. Annu Rev Astron Astrophys 51:511–653. doi:10.1146/annurev-astro-082708-101811. arXiv:1304.7762

Penrose R (1959) The apparent shape of a relativistically moving sphere. Math Proc Cambridge Philos Soc 55(01):137–139. doi:10.1017/S0305004100033776

Reid MJ, Menten KM, Zheng XW, Brunthaler A, Moscadelli L, Xu Y, Zhang B, Sato M, Honma M, Hirota T, Hachisuka K, Choi YK, Moellenbrock GA, Bartkiewicz A (2009) Trigonometric parallaxes of massive star-forming regions. VI. Galactic structure, fundamental parameters, and noncircular motions. Astrophys J 700(1):137–148. doi:10.1088/0004-637X/700/1/137

Synge JL (1966) The escape of photons from gravitationally intense stars. Mon Notices Royal Astron Soc 131:463–466. http://dx.doi.org/10.1093/mnras/131.3.463

Walsh JL, Barth AJ, Ho LC, Sarzi M (2013) The M87 black hole mass from gas-dynamical models of space telescope imaging spectrograph observations. Astrophys J 770(2):86(11). doi:10.1088/0004-637X/770/2/86

Chapter 5
Conclusions

Abstract This chapter contains a summary of important properties of the shadow of black holes and an outlook on upcoming extension.

Based on a detailed analysis of the Plebański–Demiański metric of black hole space-times, we have derived analytical formulas for the photon regions and for the shadow of a black hole.[1] They are valid for all Plebański–Demiański space-times although the involved metric functions have different meanings. In general, the space-times are not asymptotically flat and may have a cosmological horizon. Therefore, one cannot restrict to observers at infinity as it is done in many articles by other authors on shadows of black holes. Our formalism allows for observers at any Boyer–Lindquist coordinates in the domain of outer communication.

It turns out that the shape of the black hole's shadow is mostly determined by the spin of the black hole where the resulting asymmetric deformation can be understood as a *dragging effect* of the rotation. But it is unexpected that besides the charge or the cosmological constant also the NUT charge apparently only affects the size of the shadow. Since a gravitomagnetic NUT charge causes a North–South violation of the symmetry of the photon region, which determines the boundary of the shadow, one would expect an influence on the symmetry of the shadow. However, the calculations show that the shadow seen by the standard observer stays symmetric with respect to a horizontal axis, even for non-vanishing NUT parameter and for an observer off the equatorial plane. Although the acceleration parameter does not destroy the symmetry of the shadow with respect to a horizontal axis, it does have such an effect on the photon region, the ergosphere and the causality violating region.

It is possible to consider observers in different states of motion. For a radial motion, the shadow changes according to Penrose's aberration formula that is easily deduced from my formulas. For other states of motion, the shadow is shifted or scaled. The shadow stays symmetric with respect to a horizontal axis as long as the observer does

[1]The conclusions in the first and fifth paragraph can be found in [1] while the last sentence of the second paragraph as well as the fourth paragraph are taken from [3].

A. Grenzebach, *The Shadow of Black Holes*,
SpringerBriefs in Physics, DOI 10.1007/978-3-319-30066-5_5

not move in ϑ direction. Indeed, the aberrational influence for attempts of observing a galactic black hole is small because of small relative velocities.

Our estimates of the angular diameters for the shadows of the black holes in the centers of our Galaxy and of M87 show that the shadows are roughly of the same size. Hence the planned observations may provide us with shadow images not only of the black hole in our Galaxy but also of that in M87. Astronomers expect to image the shadow of the black hole in the Galactic center in the near future.

Due to the cosmic censorship hypothesis (Penrose 2002), we have restricted ourselves to black-hole space-times, but a large part is valid for naked singularities, too. In particular, the characterization of the photon region by inequality (3.7) is true in general. A major difference is the fact that there is no domain of outer communication in case of a naked singularity. If present, the possible observer positions are restricted only by a cosmological horizon. The shadow of a naked singularity is drastically different from the shadow of a black hole, as was demonstrated by de Vries (2000) for the Kerr–Newman case. While for a black hole the shadow is two-dimensional (an area on the sky, bounded by a closed curve), for a naked singularity the shadow is one-dimensional (an arc on the sky).

Another obvious way to continue this work is guided by the question whether it is possible to reveal black hole parameters from an observed shadow of a black hole. For this, my analytical formulas for the boundary curve of the shadow seem to be a promising tool since they are valid for a general class of space-times. If also the aberrational effect is taken into account, the boundary curve depends on the observer's velocity and the parameters of the space-time (spin, electric and magnetic charge, NUT charge, the Manko–Ruiz parameter, acceleration and the cosmological constant; the mass m gives an overall scale). To make ones life easier, it is enough to start with the Kerr–Newman space-time as suggested by the no-hair theorem. The parameter reconstruction is one of the main goals of the BlackHoleCam or the Event Horizon Telescope projects.

Our analytic way to describe the boundary curve of the shadow is highly idealized since no effects of matter can be modeled. This has to be done with ray-tracing methods to achieve realistic images. The big advantage of our formulas is that they image the shadow of very general classes of black hole metrics. The many singularities that ray-tracing must overcome are overcome naturally. This provides a powerful tool for testing the shadow from Sgr A* or M87 against alternative models of black holes, as well as alternative theories of gravity.

By [1–3] I refer to my papers Grenzebach et al. (2014), Grenzebach (2015) and Grenzebach et al. (2015), respectively. Sentences marked with [i] can be found in total or only slightly modified in the ith paper

References

Grenzebach A, Perlick V, Lämmerzahl C (2014) Photon regions and shadows of Kerr–Newman–NUT Black Holes with a cosmological constant. Phys Rev D 89:124,004(12). doi:10.1103/PhysRevD.89.124004. arXiv:1403.5234

Grenzebach A (2015) Aberrational effects for shadows of black holes. In: Puetzfeld et al Proceedings of the 524th WE-Heraeus-Seminar "Equations of Motion in Relativistic Gravity", held in Bad Honnef, Germany, 17–23 Feb 2013, pp 823–832. doi:10.1007/978-3-319-18335-0_25, arXiv:1502.02861

Grenzebach A, Perlick V, Lämmerzahl C (2015) Photon regions and shadows of accelerated black holes. Int J Mod Phys D 24(9):1542,024(22). doi:10.1142/S0218271815420249 ("Special Issue Papers" of the "7th Black Holes Workshop", Aveiro, Portugal, arXiv:1503.03036)

Penrose R (2002) Gravitational Collapse: The Role of General Relativity. Gen Relativ Gravit 34(7):1141–1165. 1st published: Rivista del Nuovo Cimento, Numero Speziale I, 252 (1969); reprint as "Golden Oldie". doi:10.1023/A:1016578408204

de Vries A (2000) The apparent shape of a rotating charged black hole, closed photon orbits and the bifurcation set A_4. Class Quantum Gravity 17:123–144. doi:10.1088/0264-9381/17/1/309

Appendix A
Figure Parameters

Abstract The tables in this appendix list the parameters of the space-times for which the photon regions and shadows are plotted in Sects. 3 and 4.

The subsequent tables with space-time parameters for the Plebański–Demiański class are split into three blocks. The first one comprises the black hole parameters $\frac{a}{a_{\max}}$, β, ℓ, C, α, Λ and the resulting maximal spin $a_{\max}$ that characterizes the space-time. Since all values are measured in terms of the mass m of the black hole, we denote all with $m = 1$. In the second block, the roots and, if non-zero, the leading coefficient b_4 of the quartic polynomial Δ_r are listed. The sign of b_4 defines the causal character of ∂_r for large values of r while the roots represent the horizons of the space-time; furthermore, for each black hole space-time the outer event horizon r_+ is marked in **bold** so that horizons at larger values $r > r_+$ are cosmological horizons. The third block is relevant if $\Lambda \neq 0$ or $\alpha \neq 0$. Then, $\Delta_\vartheta \not\equiv 1$ and $\Delta_\vartheta > 1$ is guaranteed for $a_3^2 + 4a_4 < 0$ or if the right hand side (r.h.s.) of Eq. (2.21) is larger than 1. One of these conditions is fulfilled for all considered space-times. Table A.1 shows which tables correspond to which figures.

Table A.1 Overview of corresponding tables and figures

	Photon regions	Shadows	Space-times
Table A.2	Figs. 3.4–3.7	Fig. 4.5	K a, KN β, KNUT ℓ, KNNUT
Table A.3	Fig. 3.8		KNNUT: singularity C
Table A.4		Fig. 4.6	KN
Table A.5	Fig. 3.9	Fig. 4.8	P: cosmological constant Λ
Table A.6	Fig. 3.11	Fig. 4.9	PD: acceleration α
Table A.7	Fig. 3.12	Fig. 4.10	PD: signs of a, ℓ, α
		Fig. 4.11	Moving observer
		Fig. 4.12	Varying C
		Fig. 4.13	Inclination of observer

A. Grenzebach, *The Shadow of Black Holes*,
SpringerBriefs in Physics, DOI 10.1007/978-3-319-30066-5

Table A.2 Black hole parameters for Kerr–Newman–NUT space-times ($m = 1$)

	BH parameter							Δ_r		
	$\frac{a}{a_{\max}}$	β	ℓ	C	α	Λ	$a_{\max}$	roots		figure
Kerr	$\frac{1}{50}$	0	0	0	0	0	1	0.0002	**2.00**	3.4 – 3.6
	$\frac{2}{5}$	0	0	0	0	0	1	0.08	**1.92**	
	$\frac{4}{5}$	0	0	0	0	0	1	0.4	**1.6**	
	1	0	0	0	0	0	1	1	**1**	
Kerr–Newman	$\frac{1}{50}$	$\frac{5}{9}$	0	0	0	0	$\frac{2}{3}$	0.33	**1.67**	3.4
	$\frac{2}{5}$	$\frac{5}{9}$	0	0	0	0	$\frac{2}{3}$	0.39	**1.61**	
	$\frac{4}{5}$	$\frac{5}{9}$	0	0	0	0	$\frac{2}{3}$	0.6	**1.4**	
	1	$\frac{5}{9}$	0	0	0	0	$\frac{2}{3}$	1	**1**	
	$\frac{1}{50}$	$\frac{8}{9}$	0	0	0	0	$\frac{1}{3}$	0.67	**1.33**	
	$\frac{2}{5}$	$\frac{8}{9}$	0	0	0	0	$\frac{1}{3}$	0.70	**1.31**	
	$\frac{4}{5}$	$\frac{8}{9}$	0	0	0	0	$\frac{1}{3}$	0.8	**1.2**	
	1	$\frac{8}{9}$	0	0	0	0	$\frac{1}{3}$	1	**1**	
Kerr–NUT	$\frac{1}{50}$	0	$\frac{3}{4}$	0	0	0	$\frac{5}{4}$	−0.25	**2.25**	3.5
	$\frac{2}{5}$	0	$\frac{3}{4}$	0	0	0	$\frac{5}{4}$	−0.15	**2.15**	
	$\frac{4}{5}$	0	$\frac{3}{4}$	0	0	0	$\frac{5}{4}$	0.25	**1.75**	
	1	0	$\pm\frac{3}{4}$	0	0	0	$\frac{5}{4}$	1	**1**	3.7
	$\frac{1}{50}$	0	$\frac{4}{3}$	0	0	0	$\frac{5}{3}$	−0.67	**2.67**	
	$\frac{2}{5}$	0	$\frac{4}{3}$	0	0	0	$\frac{5}{3}$	−0.53	**2.53**	
	$\frac{4}{5}$	0	$\frac{4}{3}$	0	0	0	$\frac{5}{3}$	0	**2**	
	1	0	$\frac{4}{3}$	0	0	0	$\frac{5}{3}$	1	**1**	
Kerr–Newman–NUT	$\frac{1}{50}$	$\frac{5}{9}$	$\frac{3}{4}$	0	0	0	$\frac{\sqrt{145}}{12}$	−0.003	**2.00**	3.6
	$\frac{2}{5}$	$\frac{5}{9}$	$\frac{3}{4}$	0	0	0	$\frac{\sqrt{145}}{12}$	0.08	**1.92**	
	$\frac{4}{5}$	$\frac{5}{9}$	$\frac{3}{4}$	0	0	0	$\frac{\sqrt{145}}{12}$	0.40	**1.60**	
	1	$\frac{5}{9}$	$\pm\frac{3}{4}$	0	0	0	$\frac{\sqrt{145}}{12}$	1	**1**	3.7
	$\frac{1}{50}$	$\frac{8}{9}$	$\frac{4}{3}$	0	0	0	$\frac{\sqrt{17}}{3}$	−0.37	**2.37**	
	$\frac{2}{5}$	$\frac{8}{9}$	$\frac{4}{3}$	0	0	0	$\frac{\sqrt{17}}{3}$	−0.26	**2.26**	
	$\frac{4}{5}$	$\frac{8}{9}$	$\frac{4}{3}$	0	0	0	$\frac{\sqrt{17}}{3}$	0.18	**1.82**	
	1	$\frac{8}{9}$	$\frac{4}{3}$	0	0	0	$\frac{\sqrt{17}}{3}$	1	**1**	

Table A.3 Black hole parameters for varying singularity parameter C ($m = 1$), see Fig. 3.8 for the corresponding photon regions

BH parameter							Δ_r					Δ_ϑ
$\frac{a}{a_{\max}}$	β	ℓ	C	α	Λ	$a_{\max}$	roots				b_4	$a_3^2+4a_4$
$\frac{4}{5}$	$\frac{5}{9}$	$\frac{4}{3}$	-2	0	0	$\frac{2\sqrt{5}}{3}$		0.11	**1.89**			
$\frac{4}{5}$	$\frac{5}{9}$	$\frac{4}{3}$	-1	0	0	$\frac{2\sqrt{5}}{3}$		0.11	**1.89**			
$\frac{4}{5}$	$\frac{5}{9}$	$\frac{4}{3}$	$-\frac{1}{2}$	0	0	$\frac{2\sqrt{5}}{3}$		0.11	**1.89**			
$\frac{4}{5}$	$\frac{5}{9}$	$\frac{4}{3}$	0	0	0	$\frac{2\sqrt{5}}{3}$		0.11	**1.89**			
$\frac{4}{5}$	$\frac{5}{9}$	$\frac{4}{3}$	$\frac{1}{2}$	0	0	$\frac{2\sqrt{5}}{3}$		0.11	**1.89**			
$\frac{4}{5}$	$\frac{5}{9}$	$\frac{4}{3}$	1	0	0	$\frac{2\sqrt{5}}{3}$		0.11	**1.89**			
$\frac{4}{5}$	$\frac{5}{9}$	$\frac{4}{3}$	2	0	0	$\frac{2\sqrt{5}}{3}$		0.11	**1.89**			
$\frac{4}{5}$	$\frac{5}{9}$	$\frac{4}{3}$	-2	0	$\frac{1}{100}$	1.51	−17.93	0.13	**1.98**	15.82	$-\frac{1}{300}$	−0.02
$\frac{4}{5}$	$\frac{5}{9}$	$\frac{4}{3}$	-1	0	$\frac{1}{100}$	1.51	−17.93	0.13	**1.98**	15.82	$-\frac{1}{300}$	−0.02
$\frac{4}{5}$	$\frac{5}{9}$	$\frac{4}{3}$	$-\frac{1}{2}$	0	$\frac{1}{100}$	1.51	−17.93	0.13	**1.98**	15.82	$-\frac{1}{300}$	−0.02
$\frac{4}{5}$	$\frac{5}{9}$	$\frac{4}{3}$	0	0	$\frac{1}{100}$	1.51	−17.93	0.13	**1.98**	15.82	$-\frac{1}{300}$	−0.02
$\frac{4}{5}$	$\frac{5}{9}$	$\frac{4}{3}$	$\frac{1}{2}$	0	$\frac{1}{100}$	1.51	−17.93	0.13	**1.98**	15.82	$-\frac{1}{300}$	−0.02
$\frac{4}{5}$	$\frac{5}{9}$	$\frac{4}{3}$	1	0	$\frac{1}{100}$	1.51	−17.93	0.13	**1.98**	15.82	$-\frac{1}{300}$	−0.02
$\frac{4}{5}$	$\frac{5}{9}$	$\frac{4}{3}$	2	0	$\frac{1}{100}$	1.51	−17.93	0.13	**1.98**	15.82	$-\frac{1}{300}$	−0.02

Table A.4 Black hole parameters for Kerr–Newman space-times with fixed spin values a ($m = 1$), see Fig. 4.6 for the corresponding black hole shadows

BH parameter							Δ_r roots	
$\frac{a}{a_{\max}}$	β	ℓ	C	α	Λ	$a_{\max}$		
0	0	0	0	0	0	1	0	**2**
$\frac{1}{9}$	0	0	0	0	0	1	0.006	**1.99**
$\frac{2}{9}$	0	0	0	0	0	1	0.03	**1.98**
$\frac{1}{3}$	0	0	0	0	0	1	0.06	**1.94**
$\frac{4}{9}$	0	0	0	0	0	1	0.10	**1.90**
$\frac{5}{9}$	0	0	0	0	0	1	0.17	**1.83**
$\frac{2}{3}$	0	0	0	0	0	1	0.25	**1.75**
$\frac{7}{9}$	0	0	0	0	0	1	0.37	**1.63**
$\frac{8}{9}$	0	0	0	0	0	1	0.54	**1.46**
1	0	0	0	0	0	1	1	**1**
0	$\frac{5}{9}$	0	0	0	0	$\frac{2}{3}$	0.33	**1.67**
$\frac{1}{9}$	$\frac{5}{9}$	0	0	0	0	$\frac{2}{3}$	0.34	**1.66**
$\frac{2}{9}$	$\frac{5}{9}$	0	0	0	0	$\frac{2}{3}$	0.37	**1.63**
$\frac{1}{3}$	$\frac{5}{9}$	0	0	0	0	$\frac{2}{3}$	0.42	**1.58**
$\frac{4}{9}$	$\frac{5}{9}$	0	0	0	0	$\frac{2}{3}$	0.50	**1.50**
$\frac{5}{9}$	$\frac{5}{9}$	0	0	0	0	$\frac{2}{3}$	0.63	**1.37**
$\frac{2}{3}$	$\frac{5}{9}$	0	0	0	0	$\frac{2}{3}$	1	**1**
0	$\frac{8}{9}$	0	0	0	0	$\frac{1}{3}$	0.67	**1.33**
$\frac{1}{9}$	$\frac{8}{9}$	0	0	0	0	$\frac{1}{3}$	0.69	**1.31**
$\frac{2}{9}$	$\frac{8}{9}$	0	0	0	0	$\frac{1}{3}$	0.75	**1.25**
$\frac{1}{3}$	$\frac{8}{9}$	0	0	0	0	$\frac{1}{3}$	1	**1**
0	1	0	0	0	0	0	1	**1**

Table A.5 Black hole parameters for Plebański space-times ($m = 1$), see Fig. 3.9 for corresponding photon regions

	BH parameter							Δ_r					Δ_ϑ
	$\frac{a}{a_{\max}}$	β	ℓ	C	α	Λ	$a_{\max}$	roots				b_4	$a_3^2+4a_4$
Kerr	$\frac{1}{50}$	0	0	0	0	$\frac{1}{100}$	1.003	−18.25	0.0002	**2.03**	16.22	$-\frac{1}{300}$	-5×10^{-6}
	$\frac{2}{5}$	0	0	0	0	$\frac{1}{100}$	1.003	−18.24	0.08	**1.94**	16.22	$-\frac{1}{300}$	−0.002
	$\frac{4}{5}$	0	0	0	0	$\frac{1}{100}$	1.003	−18.24	0.40	**1.62**	16.22	$-\frac{1}{300}$	−0.009
	1	0	0	0	0	$\frac{1}{100}$	1.003	−18.24	1.01	**1.01**	16.22	$-\frac{1}{300}$	−0.01
	$\frac{1}{50}$	0	0	0	0	$\frac{6}{100}$	1.02	−7.91	0.0002	**2.22**	5.70	$-\frac{1}{50}$	-3×10^{-5}
	$\frac{2}{5}$	0	0	0	0	$\frac{6}{100}$	1.02	−7.91	0.09	**2.12**	5.71	$-\frac{1}{50}$	−0.01
	$\frac{4}{5}$	0	0	0	0	$\frac{6}{100}$	1.02	−7.91	0.42	**1.75**	5.74	$-\frac{1}{50}$	−0.05
	1	0	0	0	0	$\frac{6}{100}$	1.02	−7.90	1.07	**1.07**	5.76	$-\frac{1}{50}$	−0.08
Kerr–Newman	$\frac{1}{50}$	$\frac{5}{9}$	0	0	0	$\frac{1}{100}$	0.67	−18.26	0.33	**1.69**	16.24	$-\frac{1}{300}$	-2×10^{-6}
	$\frac{2}{5}$	$\frac{5}{9}$	0	0	0	$\frac{1}{100}$	0.67	−18.26	0.39	**1.63**	16.24	$-\frac{1}{300}$	−0.001
	$\frac{4}{5}$	$\frac{5}{9}$	0	0	0	$\frac{1}{100}$	0.67	−18.26	0.60	**1.41**	16.24	$-\frac{1}{300}$	−0.004
	1	$\frac{5}{9}$	0	0	0	$\frac{1}{100}$	0.67	−18.26	1.01	**1.01**	16.24	$-\frac{1}{300}$	−0.06
	$\frac{1}{50}$	$\frac{5}{9}$	0	0	0	$\frac{6}{100}$	0.69	−7.94	0.33	**1.81**	5.79	$-\frac{1}{50}$	-2×10^{-5}
	$\frac{2}{5}$	$\frac{5}{9}$	0	0	0	$\frac{6}{100}$	0.69	−7.94	0.39	**1.75**	5.80	$-\frac{1}{50}$	−0.006
	$\frac{4}{5}$	$\frac{5}{9}$	0	0	0	$\frac{6}{100}$	0.69	−7.94	0.62	**1.51**	5.81	$-\frac{1}{50}$	−0.02
	1	$\frac{5}{9}$	0	0	0	$\frac{6}{100}$	0.69	−7.93	1.06	**1.06**	5.82	$-\frac{1}{50}$	−0.04
Kerr–NUT	$\frac{1}{50}$	0	$\frac{3}{4}$	0	0	$\frac{1}{100}$	1.26	−18.14	−0.25	**2.31**	16.08	$-\frac{1}{300}$	-8×10^{-6}
	$\frac{2}{5}$	0	$\frac{3}{4}$	0	0	$\frac{1}{100}$	1.26	−18.14	−0.14	**2.20**	16.08	$-\frac{1}{300}$	−0.003
	$\frac{4}{5}$	0	$\frac{3}{4}$	0	0	$\frac{1}{100}$	1.26	−18.14	0.26	**1.80**	16.09	$-\frac{1}{300}$	−0.01
	1	0	$\frac{3}{4}$	0	0	$\frac{1}{100}$	1.26	−18.14	1.02	**1.02**	16.09	$-\frac{1}{300}$	−0.02
	$\frac{1}{50}$	0	$\frac{3}{4}$	0	0	$\frac{6}{100}$	1.32	−7.69	−0.24	**2.84**	5.09	$-\frac{1}{50}$	-5×10^{-5}
	$\frac{2}{5}$	0	$\frac{3}{4}$	0	0	$\frac{6}{100}$	1.32	−7.69	−0.13	**2.68**	5.13	$-\frac{1}{50}$	−0.02
	$\frac{4}{5}$	0	$\frac{3}{4}$	0	0	$\frac{6}{100}$	1.32	−7.68	0.31	**2.13**	5.23	$-\frac{1}{50}$	−0.09
	1	0	$\frac{3}{4}$	0	0	$\frac{6}{100}$	1.32	−7.67	1.19	**1.19**	5.30	$-\frac{1}{50}$	−0.13
Kerr–Newman–NUT	$\frac{1}{50}$	$\frac{5}{9}$	$\frac{3}{4}$	0	0	$\frac{1}{100}$	1.01	−18.16	−0.002	**2.05**	16.10	$-\frac{1}{300}$	-5×10^{-6}
	$\frac{2}{5}$	$\frac{5}{9}$	$\frac{3}{4}$	0	0	$\frac{1}{100}$	1.01	−18.16	0.08	**1.97**	16.11	$-\frac{1}{300}$	−0.002
	$\frac{4}{5}$	$\frac{5}{9}$	$\frac{3}{4}$	0	0	$\frac{1}{100}$	1.01	−18.16	0.41	**1.64**	16.11	$-\frac{1}{300}$	−0.01
	1	$\frac{5}{9}$	$\frac{3}{4}$	0	0	$\frac{1}{100}$	1.01	−18.15	1.02	**1.02**	16.11	$-\frac{1}{300}$	−0.01
	$\frac{1}{50}$	$\frac{5}{9}$	$\frac{3}{4}$	0	0	$\frac{6}{100}$	1.08	−7.72	0.01	**2.46**	5.26	$-\frac{1}{50}$	-3×10^{-5}
	$\frac{2}{5}$	$\frac{5}{9}$	$\frac{3}{4}$	0	0	$\frac{6}{100}$	1.08	−7.72	0.10	**2.34**	5.28	$-\frac{1}{50}$	−0.01
	$\frac{4}{5}$	$\frac{5}{9}$	$\frac{3}{4}$	0	0	$\frac{6}{100}$	1.08	−7.71	0.46	**1.92**	5.33	$-\frac{1}{50}$	−0.06
	1	$\frac{5}{9}$	$\frac{3}{4}$	0	0	$\frac{6}{100}$	1.08	−7.71	1.17	**1.17**	5.37	$-\frac{1}{50}$	−0.09

Table A.6 Black hole parameters for Plebański–Demiański space-times ($m = 1$), see Fig. 3.11 for corresponding photon regions

	BH parameter							Δ_r						Δ_ϑ	
	$\frac{a}{a_{\max}}$	β	ℓ	C	α	Λ	$a_{\max}$	roots				b_4	$a_3^2 + 4a_4$	r.h.s. of (2.21)	
Kerr	$\frac{1}{50}$	0	0	0	$\frac{1}{8}$	0	1	−8	0.0002	**2.00**	8	−0.02	0.06	4.00	39996
	$\frac{2}{5}$	0	0	0	$\frac{1}{8}$	0	1	−8	0.08	**1.92**	8	−0.02	0.05	4.17	95.83
	$\frac{4}{5}$	0	0	0	$\frac{1}{8}$	0	1	−8	0.4	**1.6**	8	−0.02	0.02	5	20
	1	0	0	0	$\frac{1}{8}$	0	1	−8	1	**1**	8	−0.02	0	8	8
	$\frac{1}{50}$	0	0	0	$\frac{1}{4}$	0	1	−4	0.0002	**2.00**	4	−0.06	0.25	2.00	19998
	$\frac{2}{5}$	0	0	0	$\frac{1}{4}$	0	1	−4	0.08	**1.92**	4	−0.06	0.21	2.08	47.91
	$\frac{4}{5}$	0	0	0	$\frac{1}{4}$	0	1	−4	0.4	**1.6**	4	−0.06	0.09	2.5	10
	1	0	0	0	$\frac{1}{4}$	0	1	−4	1	**1**	4	−0.06	0	4	4
	$\frac{1}{50}$	0	0	0	$\frac{1}{4}$	$\frac{1}{100}$	1.004	−3.93	0.0002	**2.04**	3.79	−0.07	0.25	2.00	18848
	$\frac{2}{5}$	0	0	0	$\frac{1}{4}$	$\frac{1}{100}$	1.004	−3.93	0.08	**1.95**	3.79	−0.07	0.21	2.09	45.03
	$\frac{4}{5}$	0	0	0	$\frac{1}{4}$	$\frac{1}{100}$	1.004	−3.93	0.40	**1.62**	3.80	−0.07	0.08	2.55	9.23
	1	0	0	0	$\frac{1}{4}$	$\frac{1}{100}$	1.004	−3.93	1.01	**1.01**	3.81	−0.07	−0.02	$\in \mathbb{C}$	$\in \mathbb{C}$
Kerr–Newman	$\frac{1}{50}$	$\frac{5}{9}$	0	0	$\frac{1}{8}$	0	$\frac{2}{3}$	−8	0.33	**1.67**	8	−0.02	0.03	4.80	23.99
	$\frac{2}{5}$	$\frac{5}{9}$	0	0	$\frac{1}{8}$	0	$\frac{2}{3}$	−8	0.39	**1.61**	8	−0.02	0.02	4.97	20.57
	$\frac{4}{5}$	$\frac{5}{9}$	0	0	$\frac{1}{8}$	0	$\frac{2}{3}$	−8	0.6	**1.4**	8	−0.02	0.01	5.71	13.33
	1	$\frac{5}{9}$	0	0	$\frac{1}{8}$	0	$\frac{2}{3}$	−8	1	**1**	8	−0.02	0	8	8
	$\frac{1}{50}$	$\frac{5}{9}$	0	0	$\frac{1}{4}$	0	$\frac{2}{3}$	−4	0.33	**1.67**	4	−0.06	0.11	2.40	12.00
	$\frac{2}{5}$	$\frac{5}{9}$	0	0	$\frac{1}{4}$	0	$\frac{2}{3}$	−4	0.39	**1.61**	4	−0.06	0.09	2.48	10.28
	$\frac{4}{5}$	$\frac{5}{9}$	0	0	$\frac{1}{4}$	0	$\frac{2}{3}$	−4	0.6	**1.4**	4	−0.06	0.04	2.86	6.67
	1	$\frac{5}{9}$	0	0	$\frac{1}{4}$	0	$\frac{2}{3}$	−4	1	**1**	4	−0.06	0	4	4
	$\frac{1}{50}$	$\frac{5}{9}$	0	0	$\frac{1}{4}$	$\frac{1}{100}$	0.67	−3.93	0.33	**1.69**	3.81	−0.07	0.11	2.40	11.99
	$\frac{2}{5}$	$\frac{5}{9}$	0	0	$\frac{1}{4}$	$\frac{1}{100}$	0.67	−3.93	0.39	**1.63**	3.81	−0.07	0.09	2.49	10.18
	$\frac{4}{5}$	$\frac{5}{9}$	0	0	$\frac{1}{4}$	$\frac{1}{100}$	0.67	−3.93	0.60	**1.42**	3.81	−0.07	0.04	2.90	6.41
	1	$\frac{5}{9}$	0	0	$\frac{1}{4}$	$\frac{1}{100}$	0.67	−3.93	1.01	**1.01**	3.81	−0.07	−0.01	$\in \mathbb{C}$	$\in \mathbb{C}$
Kerr–NUT	$\frac{1}{50}$	0	$\frac{3}{4}$	0	$\frac{1}{8}$	0	1.12	−0.28	**1.98**	7.77	8.25	0.02	1×10^{-4}	−974.3	101.3
	$\frac{2}{5}$	0	$\frac{3}{4}$	0	$\frac{1}{8}$	0	1.12	−0.19	**1.89**	5.83	23.32	0.01	0.03	−83.46	6.54
	$\frac{4}{5}$	0	$\frac{3}{4}$	0	$\frac{1}{8}$	0	1.12	−62.52	0.16	**1.53**	5.68	−0.003	0.02	5.97	63.63
	1	0	$\frac{3}{4}$	0	$\frac{1}{8}$	0	1.12	−28.85	0.84	**0.84**	5.77	−0.007	0	10.80	10.80
	$\frac{1}{50}$	0	$\frac{3}{4}$	0	$\frac{1}{4}$	0	1.04	−0.31	**1.79**	3.89	4.12	0.10	6×10^{-4}	−494.2	44.53
	$\frac{2}{5}$	0	$\frac{3}{4}$	0	$\frac{1}{4}$	0	1.04	−0.23	**1.71**	2.94	10.30	0.05	0.13	−38.05	3.01
	$\frac{4}{5}$	0	$\frac{3}{4}$	0	$\frac{1}{4}$	0	1.04	−53.63	0.10	**1.37**	2.83	−0.01	0.10	3.01	54.95
	1	0	$\frac{3}{4}$	0	$\frac{1}{4}$	0	1.04	−17.58	0.72	**0.72**	2.87	−0.02	0	6.09	6.09
	$\frac{1}{50}$	0	$\frac{3}{4}$	0	$\frac{1}{4}$	$\frac{1}{100}$	1.05	−0.31	**1.84**	3.15	5.12	0.10	6×10^{-4}	−503.5	44.88
	$\frac{2}{5}$	0	$\frac{3}{4}$	0	$\frac{1}{4}$	$\frac{1}{100}$	1.05	−0.23	**1.76**	2.75	11.58	0.05	0.12	−40.22	3.06
	$\frac{4}{5}$	0	$\frac{3}{4}$	0	$\frac{1}{4}$	$\frac{1}{100}$	1.05	−38.50	0.10	**1.41**	2.69	−0.01	0.08	3.19	36.10
	1	0	$\frac{3}{4}$	0	$\frac{1}{4}$	$\frac{1}{100}$	1.05	−15.71	0.74	**0.74**	2.74	−0.03	−0.02	$\in \mathbb{C}$	$\in \mathbb{C}$
Kerr–Newman–NUT	$\frac{1}{50}$	$\frac{5}{9}$	$\frac{3}{4}$	0	$\frac{1}{8}$	0	0.86	−0.07	**1.70**	7.82	8.19	0.02	4×10^{-5}	−4937	162.0
	$\frac{2}{5}$	$\frac{5}{9}$	$\frac{3}{4}$	0	$\frac{1}{8}$	0	0.86	0.01	**1.62**	6.03	16.32	0.01	0.01	9.62	3243
	$\frac{4}{5}$	$\frac{5}{9}$	$\frac{3}{4}$	0	$\frac{1}{8}$	0	0.86	0.29	**1.35**	5.66	136.8	0.002	0.01	7.69	39.90
	1	$\frac{5}{9}$	$\frac{3}{4}$	0	$\frac{1}{8}$	0	0.86	−80.96	0.82	**0.82**	5.67	−0.003	0	12.12	12.12
	$\frac{1}{50}$	$\frac{5}{9}$	$\frac{3}{4}$	0	$\frac{1}{4}$	0	0.78	−0.14	**1.52**	3.92	4.09	0.10	2×10^{-4}	−1463	78.27
	$\frac{2}{5}$	$\frac{5}{9}$	$\frac{3}{4}$	0	$\frac{1}{4}$	0	0.78	−0.06	**1.45**	3.06	7.43	0.07	0.05	−170.0	4.76
	$\frac{4}{5}$	$\frac{5}{9}$	$\frac{3}{4}$	0	$\frac{1}{4}$	0	0.78	0.20	**1.19**	2.84	31.31	0.02	0.05	4.06	29.68
	1	$\frac{5}{9}$	$\frac{3}{4}$	0	$\frac{1}{4}$	0	0.78	−137.3	0.69	**0.69**	2.83	−0.003	0	7.05	7.05
	$\frac{1}{50}$	$\frac{5}{9}$	$\frac{3}{4}$	0	$\frac{1}{4}$	$\frac{1}{100}$	0.79	−0.13	**1.55**	3.17	5.13	0.09	2×10^{-4}	−1596	79.31
	$\frac{2}{5}$	$\frac{5}{9}$	$\frac{3}{4}$	0	$\frac{1}{4}$	$\frac{1}{100}$	0.79	−0.06	**1.48**	2.85	8.28	0.06	0.04	−235.3	4.86
	$\frac{4}{5}$	$\frac{5}{9}$	$\frac{3}{4}$	0	$\frac{1}{4}$	$\frac{1}{100}$	0.79	0.21	**1.21**	2.702	43.04	0.01	0.04	4.32	23.51
	1	$\frac{5}{9}$	$\frac{3}{4}$	0	$\frac{1}{4}$	$\frac{1}{100}$	0.79	−66.20	0.70	**0.70**	2.70	−0.01	−0.01	$\in \mathbb{C}$	$\in \mathbb{C}$

Table A.7 Black hole parameters with varying signs for Plebański–Demiański space-times ($m = 1$), see Fig. 3.12 for corresponding photon regions

	BH parameter							Δ_r					Δ_ϑ		
	$\frac{a}{a_{\max}}$	β	ℓ	C	α	Λ	$a_{\max}$	roots				b_4	$a_3^2+4a_4$	r.h.s. of (2.21)	
Kerr	$\frac{1}{50}$	0	0	0	$\pm\frac{1}{8}$	0	1	−8	0.0002	**2.00**	8	−0.02	0.06	±4.00	±39996
	$\frac{2}{5}$	0	0	0	$\pm\frac{1}{8}$	0	1	−8	0.08	**1.92**	8	−0.02	0.05	±4.17	±95.83
	$\frac{4}{5}$	0	0	0	$\pm\frac{1}{8}$	0	1	−8	0.4	**1.6**	8	−0.02	0.02	±5	±20
	1	0	0	0	$\pm\frac{1}{8}$	0	1	−8	1	**1**	8	−0.02	0	±8	±8
	$\frac{1}{50}$	0	0	0	$\pm\frac{1}{4}$	0	1	−4	0.0002	**2.00**	4	−0.06	0.25	±2.00	±19998
	$\frac{2}{5}$	0	0	0	$\pm\frac{1}{4}$	0	1	−4	0.08	**1.92**	4	−0.06	0.21	±2.09	±47.91
	$\frac{4}{5}$	0	0	0	$\pm\frac{1}{4}$	0	1	−4	0.4	**1.6**	4	−0.06	0.09	±2.5	±10
	1	0	0	0	$\pm\frac{1}{4}$	0	1	−4	1	**1**	4	−0.06	0	±4	±4
Kerr–Newman	$\frac{1}{50}$	$\frac{5}{9}$	0	0	$\pm\frac{1}{8}$	0	$\frac{2}{3}$	−8	0.33	**1.67**	8	−0.02	0.03	±4.80	±23.99
	$\frac{2}{5}$	$\frac{5}{9}$	0	0	$\pm\frac{1}{8}$	0	$\frac{2}{3}$	−8	0.39	**1.61**	8	−0.02	0.02	±4.97	±20.57
	$\frac{4}{5}$	$\frac{5}{9}$	0	0	$\pm\frac{1}{8}$	0	$\frac{2}{3}$	−8	0.6	**1.4**	8	−0.02	0.01	±5.71	±13.33
	1	$\frac{5}{9}$	0	0	$\pm\frac{1}{8}$	0	$\frac{2}{3}$	−8	1	**1**	8	−0.02	0	±8	±8
	$\frac{1}{50}$	$\frac{5}{9}$	0	0	$\pm\frac{1}{4}$	0	$\frac{2}{3}$	−4	0.33	**1.67**	4	−0.06	0.11	±2.40	±12.00
	$\frac{2}{5}$	$\frac{5}{9}$	0	0	$\pm\frac{1}{4}$	0	$\frac{2}{3}$	−4	0.39	**1.61**	4	−0.06	0.09	±2.48	±10.28
	$\frac{4}{5}$	$\frac{5}{9}$	0	0	$\pm\frac{1}{4}$	0	$\frac{2}{3}$	−4	0.6	**1.4**	4	−0.06	0.04	±2.86	±6.67
	1	$\frac{5}{9}$	0	0	$\pm\frac{1}{4}$	0	$\frac{2}{3}$	−4	1	**1**	4	−0.06	0	±4	±4
Kerr–NUT	$\frac{1}{50}$	0	$\pm\frac{3}{4}$	0	$\pm\frac{1}{8}$	0	1.12	−0.28	**1.98**	7.77	8.25	0.02	0.0001	±101.3	∓974.3
	$\frac{1}{50}$	0	$\pm\frac{3}{4}$	0	$\mp\frac{1}{8}$	0	1.45	−8.33	−7.71	−0.21	**2.67**	0.01	0.0001	±960.3	∓103.9
	$\frac{2}{5}$	0	$\pm\frac{3}{4}$	0	$\pm\frac{1}{8}$	0	1.12	−0.19	**1.89**	5.83	23.32	0.01	0.03	±6.54	∓83.46
	$\frac{2}{5}$	0	$\pm\frac{3}{4}$	0	$\mp\frac{1}{8}$	0	1.45	−44.04	−5.70	−0.09	**2.52**	0.003	0.03	±143.4	∓6.49
	$\frac{4}{5}$	0	$\pm\frac{3}{4}$	0	$\pm\frac{1}{8}$	0	1.12	−62.52	0.16	**1.53**	5.68	−0.003	0.02	±5.97	±63.63
	$\frac{4}{5}$	0	$\pm\frac{3}{4}$	0	$\mp\frac{1}{8}$	0	1.45	−5.78	0.37	**2.07**	27.15	−0.005	0.02	∓5.25	∓26.18
	1	0	$\pm\frac{3}{4}$	0	$\pm\frac{1}{8}$	0	1.12	−28.85	0.84	**0.84**	5.77	−0.007	0	±10.80	±10.80
	1	0	$\pm\frac{3}{4}$	0	$\mp\frac{1}{8}$	0	1.45	−5.93	1.24	**1.24**	18.74	−0.008	0	∓7.82	∓7.82
	$\frac{1}{50}$	0	$\pm\frac{3}{4}$	0	$\pm\frac{1}{4}$	0	1.04	−0.31	**1.79**	3.89	4.12	0.10	0.0006	±44.54	∓494.2
	$\frac{1}{50}$	0	$\pm\frac{3}{4}$	0	$\mp\frac{1}{4}$	0	1.75	−4.20	−3.83	−0.16	**3.46**	0.03	0.0006	∓46.22	±507.0
	$\frac{2}{5}$	0	$\pm\frac{3}{4}$	0	$\pm\frac{1}{4}$	0	1.04	−0.23	**1.71**	2.94	10.30	0.05	0.13	±3.01	∓38.05
	$\frac{2}{5}$	0	$\pm\frac{3}{4}$	0	$\mp\frac{1}{4}$	0	1.75	−82.69	−2.83	−0.02	**3.13**	0.003	0.12	∓2.95	±253.4
	$\frac{4}{5}$	0	$\pm\frac{3}{4}$	0	$\pm\frac{1}{4}$	0	1.04	−53.63	0.10	**1.37**	2.83	−0.01	0.10	±3.01	±55.0
	$\frac{4}{5}$	0	$\pm\frac{3}{4}$	0	$\mp\frac{1}{4}$	0	1.75	−2.95	0.55	**2.55**	9.77	−0.03	0.10	∓2.31	∓8.81
	1	0	$\pm\frac{3}{4}$	0	$\pm\frac{1}{4}$	0	1.04	−17.58	0.72	**0.72**	2.87	−0.02	0	±6.09	±6.09
	1	0	$\pm\frac{3}{4}$	0	$\mp\frac{1}{4}$	0	1.75	−3.05	1.58	**1.58**	7.61	−0.03	0	∓3.18	∓3.18

Appendix B
Isometries

Abstract In this appendix, it is shown that the coordinate transformations given in Sect. 2.2 are indeed isomorphisms within the Plebański–Demiański class.

Definition A map $(M, g) \rightarrow (M', g')$ between semi-Riemannian manifolds is an *isometry* iff there is a coordinate transformation f such that

- $f: M \rightarrow M'$ is a diffeomorphism,
- $g = f^* g'$, i.e., g is the pullback of g'.

In Riemannian geometry, an isometry is thus a metric conserving map which means that the length of vectors or curves is preserved but possibly not the distance between two points. This is different compared to isometries between ordinary metric spaces which also preserve distances.

We show that the coordinate transformations given in Eq. (2.11)

$$\begin{aligned} f^1 \colon (M_{[m,a,\beta,\ell,C,\alpha,\lambda]}, g) &\longrightarrow (M'_{[m,-a,\beta,\ell,-C,\alpha,\lambda]}, g') \\ (r, \vartheta, \varphi, t) &\longmapsto (r', \vartheta', \varphi', t') = (r, \pi - \vartheta, -\varphi, t) \end{aligned} \tag{B.1}$$

$$\begin{aligned} f^2 \colon (M_{[m,a,\beta,\ell,C,\alpha,\lambda]}, g) &\longrightarrow (M'_{[m,a,\beta,-\ell,-C,-\alpha,\lambda]}, g') \\ (r, \vartheta, \varphi, t) &\longmapsto (r', \vartheta', \varphi', t') = (r, \pi - \vartheta, \varphi, t) \end{aligned} \tag{B.2}$$

are indeed isometries between two space-times of the Plebański–Demiański class.

It is clear that the f^i are diffeomorphisms. Because the metric coefficients (2.1) are independent of φ, they are invariant under $\varphi \mapsto -\varphi$. Thus, we only have to consider $\vartheta \mapsto \pi - \vartheta$. For both transformations (B.1) and (B.2), we find

$$\Omega' = 1 - \tfrac{\alpha'}{\omega'}(\ell' - a' \cos \vartheta) r = \Omega, \tag{B.3a}$$

$$\Sigma' = r^2 + (\ell' - a' \cos \vartheta)^2 = \Sigma, \tag{B.3b}$$

$$\chi' = a' \sin^2 \vartheta - 2\ell'(-\cos \vartheta + C') = \begin{cases} -\chi & \text{if } a' = -a, \\ \chi & \text{if } {}^{\ell' = -\ell}_{\alpha' = -\alpha}, \end{cases} \tag{B.3c}$$

A. Grenzebach, *The Shadow of Black Holes*,
SpringerBriefs in Physics, DOI 10.1007/978-3-319-30066-5

$$\Delta'_{\vartheta'} = 1 + a'_3 \cos\vartheta - a'_4 \cos^2\vartheta = \Delta_\vartheta, \tag{B.3d}$$

$$\Delta'_{r'} = b'_0 + b'_1 r + b'_2 r^2 + b'_3 r^3 + b'_4 r^4 = \Delta_r. \tag{B.3e}$$

From (2.3)–(2.5) one can read off that the changed sign for a or ℓ and α yields $a'_3 = -a_3$, $a'_4 = a_4$ and $b'_j = b_j$ which verifies (B.3d) and (B.3e). Thus

$$g'_{r'r'} = g_{rr}, \qquad g'_{\varphi'\varphi'} = g_{\varphi\varphi}, \qquad g'_{\varphi' t'} = \begin{cases} -g_{\varphi t} & \text{if } a' = -a, \\ g_{\varphi t} & \text{if } {\ell' = -\ell \atop \alpha' = -\alpha}. \end{cases}$$
$$g'_{\vartheta'\vartheta'} = g_{\vartheta\vartheta}, \qquad g'_{t't'} = g_{tt}, \tag{B.4}$$

For the bases $\left\{\frac{\partial}{\partial r}, \frac{\partial}{\partial\vartheta}, \frac{\partial}{\partial\varphi}, \frac{\partial}{\partial t}\right\} =: \left\{\frac{\partial}{\partial x^i}\right\}$ and $\left\{\frac{\partial}{\partial r'}, \frac{\partial}{\partial\vartheta'}, \frac{\partial}{\partial\varphi'}, \frac{\partial}{\partial t'}\right\} =: \left\{\frac{\partial}{\partial y^j}\right\}$ of the tangent spaces TM and TM', respectively, the pushforward of $v \in TM$ belonging to $f = (f_r, f_\vartheta, f_\varphi, f_t)\colon M \to M'$ is

$$f_*(v) = \mathrm{d}f(v) = \sum_i v_i \,\mathrm{d}f\left(\tfrac{\partial}{\partial x^i}\right) = \sum_i v_i \sum_j \frac{\partial f_j}{\partial x^i}\frac{\partial}{\partial y^j} = \sum_j \left(\sum_i \frac{\partial f_j}{\partial x^i} v_i\right)\frac{\partial}{\partial y^j} \tag{B.5}$$

Applied to f^i we find with respect to the basis $\left\{\frac{\partial}{\partial y^j}\right\}$

$$f^1_*(v) = (v_r, -v_\vartheta, -v_\varphi, v_t), \qquad f^2_*(v) = (v_r, -v_\vartheta, v_\varphi, v_t). \tag{B.6}$$

This yields for the pullback of g'

$$\begin{aligned} f^{1*}g'(v,w) &= g'(f^1_* v, f^1_* w) \\ &= v \begin{bmatrix} 1 & & & \\ & -1 & & \\ & & -1 & \\ & & & 1 \end{bmatrix} \begin{bmatrix} g_{rr} & & & \\ & g_{\vartheta\vartheta} & & \\ & & g_{\varphi\varphi} & -g_{\varphi t} \\ & & -g_{\varphi t} & g_{tt} \end{bmatrix} \begin{bmatrix} 1 & & & \\ & -1 & & \\ & & -1 & \\ & & & 1 \end{bmatrix} w^t \\ &= v \begin{bmatrix} g_{rr} & & & \\ & g_{\vartheta\vartheta} & & \\ & & g_{\varphi\varphi} & g_{\varphi t} \\ & & g_{\varphi t} & g_{tt} \end{bmatrix} w^t = g(v,w), \end{aligned} \tag{B.7}$$

$$f^{2*}g'(v,w) = g'(f^2_* v, f^2_* w) = \cdots = g(v,w). \tag{B.8}$$

Hence f^1 and f^2 are isometries. ■

Printed in the United States
By Bookmasters